T0253918

A Student's Guide to the Mathematics of Astronomy

The study of astronomy offers an unlimited opportunity for us to gain a deeper understanding of our planet, the Solar System, the Milky Way galaxy, and the known Universe.

Using the plain-language approach that has proven highly popular in Fleisch's other *Student's Guides*, this book is ideal for non-science majors taking introductory astronomy courses. The authors address topics that students find most troublesome, on subjects ranging from stars and light to gravity and black holes. Dozens of fully worked examples and over 150 exercises and homework problems help readers get to grips with the concepts presented in each chapter.

An accompanying website, available at www.cambridge.org/9781107610217, features a host of supporting materials, including interactive solutions for every exercise and problem in the text and a series of video podcasts in which the authors explain the important concepts of every section of the book.

DANIEL FLEISCH is a Professor in the Department of Physics at Wittenberg University, Ohio, where he specializes in electromagnetics and space physics. He is the author of *A Student's Guide to Maxwell's Equations* and *A Student's Guide to Vectors and Tensors* (Cambridge University Press 2008 and 2011, respectively).

JULIA KREGENOW is a Lecturer in Astronomy at the Pennsylvania State University, where she is involved in researching how to more effectively teach science to non-science majors.

A Student's Guide to the Mathematics of Astronomy

Daniel Fleisch
Wittenberg University

Julia Kregenow
Pennsylvania State University

CAMBRIDGE
UNIVERSITY PRESS

CAMBRIDGE
UNIVERSITY PRESS

Shaftesbury Road, Cambridge CB2 8EA, United Kingdom

One Liberty Plaza, 20th Floor, New York, NY 10006, USA

477 Williamstown Road, Port Melbourne, VIC 3207, Australia

314–321, 3rd Floor, Plot 3, Splendor Forum, Jasola District Centre, New Delhi – 110025, India

103 Penang Road, #05–06/07, Visioncrest Commercial, Singapore 238467

Cambridge University Press is part of Cambridge University Press & Assessment, a department of the University of Cambridge.

We share the University's mission to contribute to society through the pursuit of education, learning and research at the highest international levels of excellence.

www.cambridge.org
Information on this title: www.cambridge.org/9781107610217

© D. Fleisch and J. Kregenow 2013

First published 2013
5th printing 2018

A catalogue record for this publication is available from the British Library

Library of Congress Cataloging-in-Publication data
Fleisch, Daniel A.
A student's guide to the mathematics of astronomy / Daniel Fleisch and Julia Kregenow.
pages cm.
ISBN 978-1-107-61021-7 (pbk.)
1. Astronomy – Mathematics – Textbooks. I. Kregenow, Julia. II. Title.
QB51.3.M38F54 2013
520.1′51–dc23
2013008432

ISBN 978-1-107-03494-5 Hardback
ISBN 978-1-107-61021-7 Paperback

Additional resources for this publication at www.cambridge.org/9781107610217

Contents

Preface

This book has one purpose: to help you understand and apply the mathematics used in college-level astronomy. The authors have instructed several thousand students in introductory astronomy courses at large and small universities, and in our experience a common response to the question "How's the course going for you?" is "I'm doing fine with the concepts, but I'm struggling with the math." If you're a student in that situation, or if you're a life-long learner who'd like to be able to delve more deeply into the many wonderful astronomy books and articles in bookstores and on-line, this book is here to help.

We want to be clear that this book is not intended to be your first exposure to astronomy, and it is not a comprehensive treatment of the many topics you can find in traditional astronomy textbooks. Instead, it provides a detailed treatment of selected topics that our students have found to be mathematically challenging. We have endeavored to provide just enough context for those topics to help foster deeper understanding, to explain the meaning of important mathematical relationships, and most of all to provide lots of illustrative examples.

We've also tried to design this book in a way that supports its use as a supplemental text. You'll notice that the format is modular, so you can go right to the topic of interest. If you're solid on gravity but uncertain of how to use the radiation laws, you can skip Chapter 2 and dive right into Section 3.2 of Chapter 3. Additionally, we've put a detailed discussion of four foundational topics right up front in Chapter 1, so you can work through those if you're in need of some review on unit conversions, using ratios, rate problems, or scientific notation.

To help you use this book actively (rather than just passively reading the words), we've put one or more exercises at the end of most subsections. These exercises are usually drills of a single concept or mathematical operation just discussed, and you'll find a full solution to every exercise on the book's website. Additionally, at the end of each chapter you'll find approximately 10 problems. These chapter-end problems are generally more comprehensive and challenging than the exercises, often requiring you to synthesize multiple concepts and techniques to find the solution. Full solutions for all problems are available on the book's website, and those solutions are interactive. That means you'll be able to view the entire solution straightaway, or you can request a hint to help you get started. Then, as you work through the problem, if you get stuck you can ask for additional hints (one at a time) until you finally reach the full solution.

Another resource on the book's website is a series of video podcasts in which we work through each section of the book, discussing important concepts and techniques and providing additional explanations of equations and graphs. In keeping with the modular nature of the book, we've made these podcasts as stand-alone as possible, so you can watch them all in order, or you can skip around and watch only those podcasts on the topics with which you need help.

The book's website also provides links to helpful resources for topics such as the nature of light, the center of mass, conic sections, potential energy, and significant figures (so you'll know when you should keep lots of decimal places and when it's safe to round your results).

So if you're interested in astronomy and have found mathematics to be a barrier to your learning, we're here to help. We hope this book and the supporting materials will help you turn that barrier into a stepping stool to reach a higher level of understanding. Whether you're a college student seeking additional help with the mathematics of your astronomy course or a life-long learner working on your own, we commend your initiative.

Acknowledgements

This book grew out of conversations and help sessions with many astronomy students over the years. The initiative of those students in asking thoughtful questions, often in the face of deep-seated math anxiety, inspired us not only to write this book, but to make every explanation as clear and complete as possible. In addition to inspiration, our students have provided detailed feedback as to which topics are most troublesome and which explanations are most helpful, and those are the topics and explanations that appear in this book. For this inspiration and guidance, we thank our students.

Julia also thanks Jason Wright for his moral support throughout the project and for sharing his technical expertise on stars, and she thanks Mel Zernow for his helpful comments on an early draft.

Dan thanks Gracie Winzeler for proving that every math problem can be overcome by persistence and determination. And as always, Dan cannot find the words to properly express his gratitude to the galactically terrific Jill Gianola.

1

Fundamentals

This chapter reviews four important mathematical concepts and techniques that will be helpful in many quantitative problems you're likely to encounter in a college-level introductory astronomy course or textbook. As with all the chapters in the book, you can read the sections within this chapter in any order, or you can skip them entirely if you're already comfortable with this material. But if you're working through one of the later chapters and you find that you're uncertain about some aspect of unit conversion, the ratio method, rate problems, or scientific notation, you can turn back to the relevant section of this chapter.

1.1 Units and unit conversions

One of the most powerful tools you can use in solving problems and in checking your solutions is to consistently include *units* in your calculations. As you may have noticed, among the first things that physics and astronomy professors look for when checking students' work is whether the units of the answer make sense. Students who become adept at problem-solving develop the habit of checking this for themselves.

Understanding units is important not just in science, but in everyday life as well. That's because units are all around you, giving meaning to the numbers that precede them. Telling someone "I have a dozen" is meaningless. A dozen what? Bagels? Minutes to live? Spouses? If you hope to communicate information about quantities to others, numbers alone are insufficient. Nearly every number must have units to define its meaning. So a very good habit to start building mastery is to always include the units of any number you write down.

Of course, some numbers are inherently "unitless." As an example of such a number, consider what happens when you divide the mass of the Sun

$(2 \times 10^{30}$ kg) by the mass of the Earth $(6 \times 10^{24}$ kg) in order to compare their values. The result of this division is approximately 333,333. Not 333,333 kg, just 333,333, because the units of kilograms in the numerator and denominator cancel, as explained later in this section. This unit cancellation happens whenever you divide two numbers with the same units, so you'll see several unitless numbers in Section 1.2 of this chapter.

If keeping track of units is the vital first step in solving astronomy problems, knowing how to reliably convert between different units is a close second. When you travel to a country that uses a different currency, you learn firsthand the importance of unit conversions. If you come upon a restaurant offering a full dinner for 500 rupees, is that a good deal? You'll have to do a unit conversion to find out. And to do that conversion, you'll need two things: (1) a conversion factor between currencies, such as those shown in Figure 1.1; and (2) knowledge of how to use conversion factors.

To understand the process of unit conversion, it's best to start with simple cases using everyday units, because you probably have an intuitive sense of how to perform such conversions. For example, if a movie lasts 2 hours, you know that is 120 minutes, because there are 60 minutes in 1 hour. But think about the process you used to convert hours to minutes: you intuitively multiplied 2 hours by 60 minutes in each hour.

Unfortunately, unit conversion becomes less intuitive when you're using units that are less familiar to you, or when you're using large numbers that can't be multiplied in your head. In such cases, students sometimes resort to guessing whether to multiply or divide the original quantity by the conversion factor. After a short discussion of conversion factors, we'll show you a foolproof method for setting up any unit conversion problem that will ensure you always know whether to multiply or divide.

1.1.1 Conversion factors

So exactly what *is* a conversion factor? It's just a statement of the equivalence between expressions with different units, and that statement lets you translate between those units in either direction. How can two expressions with different numbers be equivalent? Well, the distance represented by 1 meter is exactly the same as the distance represented by 100 cm. So it's the *underlying quantity* that's the same, and that quantity is represented by the *combination* of the number and the unit.

This means that a conversion factor is always a statement that some number of one unit is equivalent to a different number of another unit. Conversion factors are usually written in one of two ways: either as an equivalence relation

Figure 1.1 Currency exchange rates on a bank board. Each entry is a conversion factor between one unit and another.

or as a fraction. For example, 12 inches of length is equivalent to 1 foot, 60 minutes of time is equivalent to 1 hour, and the astronomical distance unit of 1 parsec (pc) is equivalent to 3.26 light years (ly). Each of these conversion factors can be expressed in an equivalence relation, which we signify using a double-headed arrow ($\leftrightarrow$):

$$12 \text{ in} \leftrightarrow 1 \text{ ft}, \qquad 1 \text{ hr} \leftrightarrow 60 \text{ min}, \qquad 3.26 \text{ ly} \leftrightarrow 1 \text{ pc}.$$

For convenience, one of the numbers in a conversion factor is often chosen to be 1, but it doesn't have to be. For example, 36 inches $\leftrightarrow$ 3 feet is a perfectly valid conversion factor.

It is convenient to represent the conversion factor as a fraction, with one set of units and its corresponding number in the numerator, and the other set in the denominator. Representing the example conversion factors shown above as fractions, you have

$$\frac{12 \text{ in}}{1 \text{ ft}} \text{ or } \frac{1 \text{ ft}}{12 \text{ in}}, \qquad \frac{60 \text{ min}}{1 \text{ hr}} \text{ or } \frac{1 \text{ hr}}{60 \text{ min}}, \qquad \frac{3.26 \text{ ly}}{1 \text{ pc}} \text{ or } \frac{1 \text{ pc}}{3.26 \text{ ly}}.$$

Because the two quantities in the conversion factor must represent the same amount, representing them as a fraction creates a numerator and a denominator that are equivalent, and thus the intrinsic *value* of the fraction is 1. You can multiply other values by this fraction with impunity, since multiplying any quantity by 1 does not change the amount – but it does change the way it looks. This is the goal of unit conversion: to change the units in which a quantity is expressed while retaining the underlying physical quantity.

Exercise 1.1. Write the following equivalence relations as fractional conversion factors:
1 in ↔ 2.54 cm,   1.6 km ↔ 1 mile,   60 arcmin ↔ 3,600 arcsec.

1.1.2 Setting up a conversion problem

The previous section explains *why* unit conversion works; here's a foolproof way to do it:

- Find the conversion factor that contains both units – the units you're given and the units to which you wish to convert.
- Write the expression you're given in the original units followed by a × symbol followed by the relevant conversion factor in fractional form.
- Multiply all the numbers and all the units of the original expression by the numbers and the units of the conversion factor. Grouping numbers and terms allows you to treat them separately, making this step easier.

You can see this method in action in the following example.

Example: Convert 1,000 minutes to hours.

The fractional forms of the relevant conversion factor (that is, the conversion factor containing hours and minutes) are $\frac{1 \text{ hr}}{60 \text{ min}}$ and $\frac{60 \text{ min}}{1 \text{ hr}}$. But how do you know which of these to use? Both are proper conversion factors, but one will help you solve this problem more directly.

To select the correct form of the conversion factor, look at the original units you're given. If those units are standing alone (as are the units of minutes in the expression "1,000 minutes"), use the conversion factor with the units you're trying to get rid of in the denominator and the units that you're trying to obtain in the numerator. That way, when you multiply, the units you don't want will cancel, and the units you want will remain. This works because you can cancel units that appear in both the numerator and the denominator of a fraction in the same way you can cancel numerical factors.

In this example, since the units you're given (minutes) appear standing alone and you want to convert to units of hours, the correct form of the conversion factor has minutes in the denominator and hours in the numerator. That factor is $\frac{1 \text{ hr}}{60 \text{ min}}$. With that conversion factor in hand, you're ready to write down the given quantity in the original units and multiply by the conversion factor. Here's how that looks with the $\boxed{\text{conversion factor}}$ boxed:

$$1{,}000 \text{ min} \times \boxed{\frac{1 \text{ hr}}{60 \text{ min}}}.$$

To simplify this expression, it helps to realize that there is an implicit multiplication between each number and its unit; and to remember that multiplication is commutative – so you can rearrange the order of the terms in both the numerator and denominator. That lets you multiply the numerical parts together and the units together, canceling units that appear on both top and bottom. Then you can simplify the numbers and express your answer in whatever units remain uncanceled:

$$1{,}000 \text{ min} \times \boxed{\frac{1 \text{ hr}}{60 \text{ min}}} = \frac{(1{,}000 \times 1)(\cancel{\text{min}} \times \text{hr})}{60 \, \cancel{\text{min}}} = \frac{1{,}000 \text{ hr}}{60} = 16.7 \text{ hr}$$

So a time value of 1,000 minutes represents the same amount of time as 16.7 hours.

Here's another example that uses the common astronomical distance units of parsecs and light years:

Example: Convert 1.29 parsecs, the distance of the closest star beyond our Sun, to light years.

In most astronomy texts, you'll find the conversion factor between parsecs and light years given as 3.26 ly $\leftrightarrow$ 1 pc, or equivalently 0.3067 pc $\leftrightarrow$ 1 ly.

In this case, since the quantity you're given has units of parsecs standing alone, you'll need the fractional conversion factor with parsecs in the denominator and light years in the numerator. Using that factor, your multiplication should look like this, again with the conversion factor boxed:

$$1.29 \text{ pc} = 1.29 \text{ pc} \times \boxed{\frac{3.26 \text{ ly}}{1 \text{ pc}}} = \frac{(1.29 \times 3.26)(\cancel{\text{pc}} \times \text{ly})}{1 \, \cancel{\text{pc}}} = \frac{4.21 \text{ ly}}{1} = 4.21 \text{ ly}.$$

Notice that the original quantity of 1.29 pc may be written as the fraction $\frac{1.29 \text{ pc}}{1}$ in order to remind you to multiply quantities in both the numerator and in the denominator. The result of this unit conversion tells you that 4.21 light years represent the same amount of distance as 1.29 parsecs. Thus, the light from the

nearest star beyond the Sun (a star called Proxima Centauri) takes over 4 years
to travel to Earth.

An additional benefit of this method of unit conversion is that it helps you
catch mistakes. Consider what would happen if you mistakenly used the conversion factor upside-down; the units of your answer wouldn't make sense.
Here are incorrect setups for the previous two examples:

$$1{,}000 \text{ min} \times \frac{60 \text{ min}}{1 \text{ hr}} = \frac{(1{,}000 \times 60)(\text{min} \times \text{min})}{1 \text{ hr}}$$

$$= 60{,}000 \frac{\text{min}^2}{\text{hr}} \text{(INCORRECT)}$$

and

$$1.29 \text{ pc} \times \frac{1 \text{ pc}}{3.26 \text{ ly}} = \frac{(1.29 \times 1)(\text{pc} \times \text{pc})}{3.26 \text{ ly}} = 0.40 \frac{\text{pc}^2}{\text{ly}} \text{ (INCORRECT)}.$$

Since these units are not the units to which you're trying to convert, you know
you must have used conversion factors incorrectly.

**Exercise 1.2. Perform the following unit conversions (you can find the
relevant conversion factors in most astronomy texts or on the Internet).**

(a) **Express 12 inches in centimeters.**
(b) **Express 100 cm in inches.**
(c) **Express 380,000 km in miles (this is roughly the distance from the
Earth to the Moon).**
(d) **Express 93,000,000 miles in kilometers (this is roughly the distance
from the Earth to the Sun).**
(e) **Express 0.5 degrees in arcseconds (this is roughly the angular size of
the full Moon viewed from Earth).**

1.1.3 Checking your answer

Whenever you do a unit conversion (or other problems in astronomy, or any
other subject for that matter), you should always give your answer a sanity
check. That is, you should ask yourself "Does my answer make sense? Is it
reasonable?" For example, in the incorrect version of the conversion from minutes to hours, you can definitely tell from the numerical part of your answer
that something went wrong. After all, since 60 minutes are equivalent to 1 hour,
then for any amount of time the number of minutes must be greater than the
equivalent number of hours. So if you were to convert 1,000 minutes to hours
and obtain an answer of 60,000 hours, the number of minutes would be smaller

than the number of hours. That means these two quantities can't possibly be equivalent, which alerts you to a mistake somewhere.

Of course, if the units are outside your common experience (such as parsecs and light years in the previous example), you might not have a sense of what is or isn't reasonable. But you'll develop that sense with practice, so be sure to always take a step back from your answer to see if it makes sense. And remember that whenever you're converting to a *larger* unit (such as minutes to hours), the numerical part of the answer should get *smaller* (so that the combination of the number and the units represents the same quantity).

Exercise 1.3. How do you know that your answers to each of the unit conversion problems in the previous exercise make sense? Give a brief explanation for each.

1.1.4 Multi-step conversions

Up to this point, we've been working with quantities that have single units, such as meters, hours, or light years. But many problems in astronomy involve quantities with multiple units, such as meters per second or watts per square meter. Happily, the conversion-factor approach works just as well for multi-unit quantities.

Example: Convert from kilometers per hour to meters per second.

Since this problem statement doesn't tell you how many km/hr, you can use 1 km/hr. To convert quantities which involve two units (kilometers and hours in this case), you can use two conversion factors in immediate succession: one to convert kilometers to meters and another to convert hours to seconds. Here's how that looks:

$$\frac{1 \text{ km}}{\text{hr}} \times \boxed{\frac{1000 \text{ m}}{1 \text{ km}}} \times \boxed{\frac{1 \text{ hr}}{3600 \text{ s}}} = \frac{(1 \times 1{,}000 \times 1)(\cancel{\text{km}} \times \text{ m} \times \cancel{\text{hr}})}{(1 \times 3{,}600)(\cancel{\text{hr}} \times \cancel{\text{km}} \times \text{ s})},$$

$$1\frac{\text{km}}{\text{hr}} = \frac{1{,}000 \text{ m}}{3{,}600 \text{ s}} = 0.28 \text{ m/s}.$$

Alternatively, you could have done two separate conversions in succession, such as km/hr to km/s, and then km/s to m/s.

You may also encounter problems in which you need to break a single conversion into multiple steps. This may occur, for example, if you don't know the conversion factor directly from the given units to the desired units, but you do know the conversions for intermediate units. This is illustrated in the following example:

Example: How many seconds old were you on your first birthday?

Even if you don't know how many seconds are in a year, you can break this problem up into years to days, then days to hours, hours to minutes, and finally minutes to seconds. So to convert between years and seconds, you could do the following:

$$1 \text{ yr} \times \boxed{\frac{365 \text{ d}}{1 \text{ yr}}} \times \boxed{\frac{24 \text{ hr}}{1 \text{ d}}} \times \boxed{\frac{60 \text{ min}}{1 \text{ hr}}} \times \boxed{\frac{60 \text{ s}}{1 \text{ min}}} = \frac{(365 \times 24 \times 60 \times 60) \text{ s}}{1}$$

$$= 31{,}536{,}000 \text{ s}.$$

By determining that there are about 31.5 million seconds in a year, you've derived the conversion factor between seconds and years. With the fractional conversion factor $\frac{31{,}536{,}000 \text{ s}}{1 \text{ yr}}$ in hand you can, for example, find the number of seconds in 30 years in a single step:

$$30 \text{ yr} \times \boxed{\frac{31{,}536{,}000 \text{ s}}{1 \text{ yr}}} = \frac{30 \times 31{,}536{,}000 \text{ s}}{1} = 946{,}080{,}000 \text{ s},$$

which is just under 1 billion. This gives you a sense of how large a billion is – you've lived a million seconds when you're about 11.5 days old, but even 30 years later you still haven't lived for a billion seconds.

Exercise 1.4. Perform the following unit conversions.

(a) Convert 60 mph (miles per hour) to meters per second.

(b) Convert 1 day to seconds.

(c) Convert dollars per kilogram to cents per gram (100 cents ↔ 1 dollar).

(d) Convert 1 mile to steps, assuming 1 step ↔ 30 inches (there are 1,760 yards in 1 mile, 3 ft in 1 yard, and 12 inches in 1 ft).

1.1.5 Converting units with exponents

Sometimes when doing a unit conversion problem, you will need to convert a unit that is raised to a power. In these cases, you must be sure to raise the conversion factor to the same power, and apply that power to all numbers *and* units in the conversion factor.

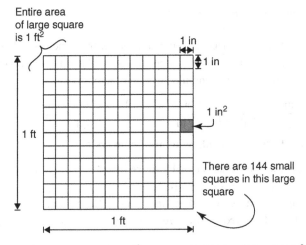

Figure 1.2 One square foot (ft^2), composed of $12\ \text{in} \times 12\ \text{in} = 144\ \text{in}^2$.

Example: Convert 1 square foot (1 ft^2) to square inches (in^2).

You already know that there are 12 inches in 1 foot. Feet and inches are both units of one-dimensional length, or *linear* dimension. *Square* feet and inches, however, are units of two-dimensional *area*. The illustration in Figure 1.2 makes it clear that one square foot is not equal to just 12 square inches, but rather 12^2, or 144 square inches.

To perform this unit conversion mathematically, without having to draw such a picture, you'd write:

$$1\ \text{ft}^2 = 1\ \text{ft}^2 \left(\frac{12\ \text{in}}{1\ \text{ft}} \right)^2 = 1\ \text{ft}^2 \left(\frac{12^2\ \text{in}^2}{1^2\ \text{ft}^2} \right) = 1\ \text{ft}^2 \left(\frac{144\ \text{in}^2}{1\ \text{ft}^2} \right) = 144\ \text{in}^2.$$

Notice that when you raise the conversion factor ($\frac{12\ \text{in}}{1\ \text{ft}}$) to the second power, both the numerical parts and the units, in both numerator and denominator, get squared.

Example: How many cubic centimeters (cm^3) are in 1 cubic meter (m^3)?

You know that there are 100 cm in 1 m, and both centimeters and meters are units of one-dimensional length. A cubic length unit, however, is a unit of three-dimensional volume. When you multiply by the appropriate conversion factor that converts between centimeters and meters, you must raise that factor to the third power, applying that power to all numbers and units separately.

$$1\ \text{m}^3 = 1\ \text{m}^3 \left(\frac{100\ \text{cm}}{1\ \text{m}} \right)^3 = 1\ \text{m}^3 \left(\frac{100^3\ \text{cm}^3}{1^3\ \text{m}^3} \right) = 1{,}000{,}000\ \text{cm}^3,$$

so there are 1 million cubic centimeters in 1 cubic meter.

Example: Convert 9.8 m/s^2 to km/hr^2.

One conversion factor is needed to convert length from meters to kilometers, and another to convert time from seconds to hours. The time conversion factor needs to be squared, but the length conversion factor does not.

$$9.8\frac{m}{s^2} = 9.8\frac{\not{m}}{\not{s^2}}\left(\frac{1\ km}{1,000\ \not{m}}\right)\left(\frac{3,600\ \not{s}}{1\ hr}\right)^2 = \frac{9.8 \times 3,600^2\ km}{1,000\ hr^2} = 127,000\frac{km}{hr^2}.$$

Exercise 1.5. Perform the following unit conversions.

(a) **How many square feet are in 1 square inch?**
(b) **Convert 1 cubic foot to cubic inches.**
(c) **How many square centimeters are in a square meter?**
(d) **Convert 1 cubic yard to cubic feet (3 feet ↔ 1 yard).**

1.1.6 Compound units

A handful of units that you're likely to encounter in an astronomy class are actually compound units, meaning that they are combinations of more basic[1] units. For example, the force unit of newtons (N) is defined as a mass in kilograms times a distance in meters divided by the square of the time in seconds: $1\ N = 1\ kg \cdot m/s^2$. This means that wherever you see units of newtons (N), you are free to replace that unit with its equivalent, $kg \cdot m/s^2$, without changing the number in front of the unit. Put another way, you can use $1\ N \leftrightarrow 1\ kg \cdot m/s^2$ to make the conversion factor $\frac{1\ N}{1\ kg \cdot m/s^2}$ or its inverse, which you can multiply by your original quantity in order to get it into a new set of units.

The energy unit joules is another example. Energy has dimensions of force (SI units of newtons) times distance (SI units of meters), so $1\ J \leftrightarrow 1\ N \cdot m$.

As one final example of compound units, the power units of watts (W) are defined as energy (SI units of joules) per time (SI units of seconds). Therefore $1\ W \leftrightarrow 1\ J/s$.

Example: Express the compound unit watts in terms of the base units kilograms (kg), meters (m), and seconds (s).

The definition of watts is given just above: energy per unit time, with SI units of joules per second:

$$1\ W \quad \leftrightarrow \quad 1\ J/s.$$

---

[1] The *base units* you will encounter in this book are those of the International System of Units ("SI"): meters for length, kilograms for mass, seconds for time, and kelvins for temperature. Many astronomers (and some astronomy texts) use the "cgs" system in which the standard units are centimeters for length, grams for mass, and seconds for time.

But joules are compound units as well:

$$1 \text{ W} = 1 \text{ J/s} = 1 \text{ (N·m)/s},$$

and newtons are compound units:

$$1 \text{ W} = 1 \text{ N·m/s} = 1 \text{ (kg·m/s}^2)·\text{m/s}.$$

This simplifies to

$$1 \text{ W} = 1 \frac{\text{kg·m}}{\text{s}^2} \cdot \frac{\text{m}}{\text{s}} = 1 \frac{\text{kg·m}^2}{\text{s}^3}.$$

This is the expression of watts in terms of SI base units. Compound units are often more convenient to use because they keep the units simpler and more compact.

Exercise 1.6. Express the following compound units in terms of base units kilograms, meters, and seconds.

(a) **Pressure: N/m^2 (note 1 N/m^2 is defined as 1 pascal, or 1 Pa).**
(b) **Energy density: J/m^3.**

The exercises throughout the section should help you practice the individual concepts and operations needed for doing unit conversions. If you're ready for some more-challenging questions that require synthesizing tools from this and other sections, take a look at the problems at the end of this chapter and the on-line solutions.

1.2 Absolute and ratio methods

On the first day of some astronomy classes, students are surprised to learn that the use of a calculator is prohibited, or at least discouraged, by the instructor. In other astronomy classes, calculators may be encouraged or even required. So what's the best way to solve problems in astronomy?

As is often the case, there is no one way that works best for everyone. There are, however, two basic methods that you're likely to find helpful. Those two methods will be referred to in this book as the *absolute method* and the *ratio method*. And although either of these methods may be used with or without a calculator, it's a good bet that if your instructor intends for you to use only the ratio method, calculators may be prohibited or discouraged.

In this book, you'll find that both the absolute method and the ratio method are used throughout the examples and problems. That way, no matter which

type of class you're in (or which method you prefer to use), you'll be able to
see plenty of relevant examples.

So exactly what are the absolute and ratio methods? The short answer is
that the absolute method is a way to determine the absolute numeric value of a
quantity in the relevant units (such as a distance of 3 meters, time duration of 15
seconds, or mass of 2 million kilograms), and the ratio method is a way to find
the unitless relative value of a quantity (such as a distance that is twice as far,
time duration that is three times as long, or mass that is 50 times greater). Of
course, for the relative value to have meaning, you must specify the reference
quantity as well (twice as far as what, for example).

1.2.1 Absolute method

The absolute method is probably the way you first learned to solve problems:
using an equation with an "equals" sign, just get the variable you're trying to
find all by itself on the left side of the equation and then plug in the values
(with units!) on the right side of the equation. So if you're trying to find the
area (A) of a circle of given radius (R), you can use the equation

$$A = \pi R^2.$$

If the radius (R) is 2 meters, the area is

$$A = (3.1416)(2 \text{ m})^2 = 12.6 \text{ m}^2.$$

The units of the answer (square meters in this case) come directly from the
units attached to the variables on the right side of the equation. Notice that
when the radius gets squared, you have to square both the number and the unit,
so $(2 \text{ m})^2$ is 2^2 m^2, or 4 m^2.

**Exercise 1.7. Calculate the following quantities for Earth, assuming a
radius (R) of 6371 km. Be sure to include units with your answer.**

(a) **The circumference (C) of the Earth's equator ($C_{circle} = 2\pi R$).**
(b) **The surface area (SA) of Earth ($SA_{sphere} = 4\pi R^2$).**
(c) **The volume (V) of Earth ($V_{sphere} = \frac{4}{3}\pi R^3$).**

1.2.2 Comparing two quantities

In everyday life, comparisons between two quantities are usually made in two
ways: either by subtracting or by dividing the quantities. For example, if one
city is 250 km away from your location, and a second city is 750 km away from
your location, you could say that the second city is 500 km farther than the first

(since 750 km −250 km = 500 km). But you could also say that the second city is three times farther away than the first (since 750 km/250 km = 3). Both of these statements are correct, but which is more useful depends on the situation. In astronomy, the values of many quantities (such as the mass of a planet, the luminosity of a star, or the distance between galaxies) are gigantic, and subtraction of one extremely large number from another can lead to results that are uncertain and difficult to interpret. In such cases, comparison by dividing is far more useful than comparison by subtracting.

For example, saying that the distance to the star Rigel is approximately 4.4 quadrillion miles (which is 4.4 million billion miles) greater than the distance to the star Vega may be useful in some situations, but saying that Rigel is about 31 *times farther* than Vega is more helpful for giving a sense of scale (and is also easier to remember). Of course, it's always possible to convert the difference in values to the ratio of the values and vice versa, provided you have the required reference information (for example, the distance to Vega). But since it's easiest to just do one comparison rather than both, your best bet is to compare using a ratio unless explicitly instructed otherwise.

It's the utility of this "comparing by dividing" idea that makes the ratio method so useful in astronomy.

Example: Compare the area of the circle you found in Section 1.2.1 (call it circle 1) to the area of another circle (call this one circle 2) with three times larger radius (so R = 6 meters for circle 2).

If you want to know how many times bigger the area of circle 2 is compared to circle 1, you could use the absolute method and calculate the area of each circle separately:

$$A_1 = \pi R_1^2 = (3.1416)(2 \text{ m})^2 = 12.6 \text{ m}^2,$$
$$A_2 = \pi R_2^2 = (3.1416)(6 \text{ m})^2 = 113.1 \text{ m}^2.$$

To compare these areas by dividing, you would then do the following

$$\frac{A_2}{A_1} = \frac{113.1 \text{ m}^2}{12.6 \text{ m}^2} = 8.98 \approx 9,$$

so the area of circle 2 is about nine times greater[2] than that of circle 1.

Notice that in addition to giving you the answer to the question "how many times bigger," this absolute method also provides the value of the area of each of the two circles (113.1 m² for circle 2 and 12.6 m² for circle 1). But if you're

---

[2] You could have taken A_1/A_2, in which case you would have gotten $\frac{A_1}{A_2} = \frac{12.6 \text{ m}^2}{113.1 \text{ m}^2} = \frac{1}{8.98} \approx \frac{1}{9}$, which is an equivalent result.

only interested in *comparing* these areas, the ratio approach can give you the answer much more quickly and easily.

Exercise 1.8. Compare the two numbers in each of the following situations using both methods of subtraction and division. Use your results to decide which method is useful in that situation, or if both might be useful.

(a) **The tallest building in Malaysia is 452 m tall. A typical person is about 1.7 m tall. How much taller is the tall building than a person?**

(b) **A man weighed 220 lb. After dieting, his weight dropped to 195 lb. How much more did he weigh before he lost the weight?**

(c) **A typical globular cluster of stars might have 400,000 stars. A typical galaxy might have 200,000,000,000 stars. How many more stars are in the galaxy?**

1.2.3 Ratio method

To understand how comparing with ratios works, try writing the equations for the areas of circles 1 and 2 from the previous example as a fraction:

$$\frac{A_2 = \pi R_2^2}{A_1 = \pi R_1^2},$$

(1.1)

which is

$$\frac{A_2}{A_1} = \frac{\not\pi R_2^2}{\not\pi R_1^2} = \frac{R_2^2}{R_1^2}$$

or

$$\frac{A_2}{A_1} = \left(\frac{R_2}{R_1}\right)^2.$$

(1.2)

Look at the simplicity of this last result: to know the ratio of area A_2 to area A_1, simply find the ratio of radius R_2 to radius R_1 and then square that value. Since you know that R_2 is three times larger than R_1, the ratio of the areas (A_2/A_1) must be nine (because $3^2 = 9$). Notice that this was the same result obtained in the previous section using the absolute method, without going through the steps necessary to individually determine the values of A_2 and A_1 and then dividing those values. The ratio method also gave you the exact answer of 9, instead of the approximate answer obtained by rounding the value of π and the values of the areas before dividing them.

Of course, in this example, those extra steps were fairly simple (squaring each radius and multiplying by π), so using the ratio method saved you only three steps – a small amount of work. But in other problems using ratios may save you many steps, so we strongly encourage you to use the ratio method whenever possible. Remember that minimizing the number of calculations you do when working a problem reduces the opportunities for errors.

Example: Compare the volumes of two spheres, one of which has three times the radius of the other.

As you may recall, the volume (V) of a sphere can be found from the sphere's radius (R) using the equation

$$V = \frac{4}{3}\pi R^3$$

in which the volume comes out in units of cubic meters (m^3) if the radius has units of meters. If you know the radius of each sphere, you could use the absolute method to find the volume of each, and then by dividing the larger volume by the smaller you could specify how many times larger that volume is. But it is preferable to use the ratio method as in the previous example:

$$\frac{V_2 = \frac{4}{3}\pi R_2^3}{V_1 = \frac{4}{3}\pi R_1^3}, \tag{1.3}$$

and, as before, all the constants in the numerator cancel with those in the denominator, leaving

$$\frac{V_2}{V_1} = \frac{R_2^3}{R_1^3}$$

or

$$\frac{V_2}{V_1} = \left(\frac{R_2}{R_1}\right)^3. \tag{1.4}$$

So to determine the ratio of the volumes you simply *cube* the ratio of the radii. You know that the larger sphere has three times the radius of the smaller, and $3^3 = 27$, so you can be certain that the larger sphere's volume is 27 times greater.

There's another way to carry through the mathematical steps to solve this type of problem. If sphere 2 has the larger radius, the relationship between the radii is $R_2 = 3R_1$. Now wherever R_2 appears in Eq. 1.4, you can replace it with $3R_1$:

$$\frac{V_2}{V_1} = \left(\frac{R_2}{R_1}\right)^3 = \left(\frac{3R_1}{R_1}\right)^3 = \left(\frac{3}{1}\right)^3 = 3^3 = 27. \tag{1.5}$$

One powerful aspect of the ratio method is that you can determine the ratio of the volumes without knowing the radius of either sphere, as long as you know the ratio of those radii. So if the large sphere has three times the radius of the small sphere its volume must be 27 times larger, irrespective of the values of the radii. The spheres could have radii of 3 meters and 9 meters, or radii of 100 meters and 300 meters, or radii of 6,000 km and 18,000 km – in every case, if the ratio of the radii is 3, the ratio of the volumes is 27.

Exercise 1.9. The Sun's radius is 109 times larger than Earth's. Use the ratio method to make the following comparisons.

(a) How many times bigger is the Sun's circumference than Earth's?
(b) How many times bigger is the Sun's surface area than Earth's?
(c) How many times bigger is the Sun's volume than Earth's?

1.2.4 Interpreting ratio answers

The last step of a ratio problem is crucial: interpreting your answer. Many students find a result but don't know what to conclude from all their work. Your result from a ratio problem typically takes the form of an equation with two variables, an equals sign, and a number, such as Eq. 1.5. You can understand your result in one of two ways. First, you can look at the ratio you were calculating (V_2/V_1) and check if its numerical equivalent $(= 27)$ is larger or smaller than 1. In this case 27 is larger than 1, so you know the quantity in the numerator (V_2) is larger than the quantity in the denominator (V_1), and by how many times (27). In other words, the volume of sphere 2 is 27 times larger than the volume of sphere 1. If the answer had been less than 1 (as would have happened if you had established sphere 1 as having the larger radius), you would have concluded the opposite, that the volume of sphere 2 was smaller, as shown in this example.

Example: You obtain the following result from a ratio problem comparing the radius of two spheres: $\frac{R_a}{R_b} = \frac{1}{5}$. Which sphere is bigger, sphere a or sphere b, and by how many times?

Inspection of the right side of the equation $\frac{R_a}{R_b} = \frac{1}{5}$ shows that the denominator (5) is larger than the numerator (1) by five times. Since this is true on the right side, it must also be true on the left, so you know that sphere b (R_b) must be larger than sphere a (R_a) by five times.

The second way to interpret a ratio answer is to make one more mathematical step of rearranging the answer, and then translate the math into words. Rearrange $\frac{V_2}{V_1} = 27$ to get $V_2 = 27 V_1$. This small equation is a mathematical sentence that conveys information. It can be mapped term by term into a complete[3] sentence in words: *"The volume of sphere 2 is 27 times the volume of sphere 1."* This gives you the physical insight you need to understand what your answer means. Understanding and being able to translate an equation into meaningful words in your native language is a very important skill.

Example: Translate the mathematical result from the previous example into words: $\frac{R_a}{R_b} = \frac{1}{5}$.

First rearrange the equation to get one variable on each side:

$$\frac{R_a}{\cancel{R_b}} \times \cancel{R_b} = \frac{1}{5} \times R_b$$

$$R_a = \frac{1}{5} R_b.$$

Now translate this term-by-term into words: "The radius of sphere a" (R_a) "is" (=) "one fifth" $\left(\frac{1}{5}\right)$ "of" ($\times$, which is implicit on the right side) "the radius of sphere b" (R_b). That is, sphere a is one-fifth as large as sphere b, so sphere *a* is smaller.

Notice that your answer would look superficially different but have the same underlying meaning if you had chosen to rearrange the equation with R_b on the left and R_a on the right. To see this, start by taking the reciprocal of both sides:

$$\left(\frac{R_a}{R_b}\right) = \frac{1}{5},$$

$$\left(\frac{R_a}{R_b}\right)^{-1} = \left(\frac{1}{5}\right)^{-1},$$

$$\left(\frac{R_b}{R_a}\right) = \frac{5}{1},$$

and then multiply both sides by R_a to get R_b by itself:

$$\left(\frac{R_b}{\cancel{R_a}}\right) \times \cancel{R_a} = \left(\frac{5}{1}\right) \times R_a,$$

$$R_b = 5 R_a.$$

Translate this term-by-term into a complete sentence of words: "The radius of sphere b is five times as large as the radius of sphere a." Saying that sphere b is five times as large as sphere a is mathematically identical to saying that sphere a is one-fifth as large as sphere b. Both phrasings make it clear that a is the smaller sphere and b is the larger, by a factor of five. So it does not matter how

---

[3] Making it a *complete* sentence ensures that you don't leave any parts out.

you choose to rearrange your result before translating, because you will get an equivalent answer.

Exercise 1.10. Interpret the following ratio results by stating which quantity is bigger and by how many times. Be sure to include a term-by-term translation of the mathematical result into complete sentences:

$$(a)\, \frac{C_a}{C_b} = \frac{1}{109}, \qquad (b)\, \frac{SA_1}{SA_2} = \frac{11,900}{1}, \qquad (c)\, \frac{V_j}{V_k} = \frac{1}{1.3 \times 10^6}.$$

1.2.5 Proportionality relationships

In many astronomy texts, the author assumes that you're going to be using the ratio method to solve at least some of the problems, so you may well encounter relationships written like this:[4]

$$C_{circle} \propto R,$$
$$A_{circle} \propto R^2,$$
$$V_{sphere} \propto R^3, \qquad\qquad (1.6)$$
$$L \propto R^2 T^4,$$

in which the $\propto$ symbol means "is proportional to." When you see such relationships, it's critical that you understand that you cannot simply replace the $\propto$ symbol with an "equals" sign. Instead, a proportionality relationship means one value equals one or more constants times the other value, where the constants are not specified. The four example expressions above come from the following equations:

$$C_{circle} = 2\pi R,$$
$$A_{circle} = \pi R^2,$$
$$V_{sphere} = \frac{4}{3}\pi R^3, \qquad\qquad (1.7)$$
$$L = 4\pi R^2 \sigma T^4.$$

Comparing Eqs. 1.6 to Eqs. 1.7, you can easily see the difference: in the proportionality relations, all of the constants (numbers or physical or mathematical constants with fixed values such as π and σ) have been left out. So the first of the relationships in Eqs. 1.6 does not say that the circumference of a circle *equals* the circle's radius, it says that the circumference *is proportional to*

---

[4] The last of these four relationships deals with the luminosity of a spherical radiating object such as a star, which you can read about in Section 3.2.2.

the radius (which means that circumference equals a constant times the radius, where the constant in this case is 2π). That's very useful if you're going to be comparing the circumferences of two circles using the ratio method, but you cannot use such a relationship to find the numerical value of a circle's circumference.

You may be wondering exactly how proportionality relations can be used in light of the fact that all the constants are omitted. You can understand the answer by considering the ratio of equations shown in Eqs. 1.1 and 1.3. In these ratios, all of the constants in the numerator are identical to those in the denominator, since the same equation underlies both, and so they cancel. And as long as all the constants are going to cancel when you make the ratio of equations, there's no need to include those constants in the first place. This means that you can use the ratio of proportionality relations in exactly the same way you used the ratio of equations.

For example, consider the ratio of proportionality relations for circles of circumference C_1 and C_2:

$$\frac{C_2 \propto R_2}{C_1 \propto R_1}.$$

Remember that the proportion symbol ("$\propto$") means "equals a constant times," where the value of the constant is not specified. When doing ratios, the constant(s) can be given any designation you choose, such as "u":

$$\frac{C_2 = uR_2}{C_1 = uR_1}.$$

The letter you use to designate the constant does not matter, because the constant cancels when a ratio is taken:

$$\frac{C_2}{C_1} = \frac{uR_2}{uR_1} = \frac{\cancel{u}R_2}{\cancel{u}R_1},$$
$$\frac{C_2}{C_1} = \frac{R_2}{R_1},$$

which means that the ratio of circumferences of two circles is equal to the ratio of their radii. So if you double the radius of a circle, the circumference also doubles. This is an example of *direct* proportionality: when one quantity gets larger by some factor, the other quantity also gets larger by that same factor. Similarly, if one quantity gets smaller by some factor, the other quantity gets smaller by that same factor.

In astronomy you will see many examples of proportional relationships in which one or more of the variables is raised to a power, such as the "R" term in the last three examples in Eqs. 1.6. In such cases, it is not correct to say that the quantity on the left side of the $\propto$ symbol is proportional to R. Instead, you should say "The area of a circle is proportional to R^2," which means that if you double the radius of a circle, the area does not just double, it increases by a factor of four (since $2^2 = 4$). Likewise, the correct statement relating a sphere's volume to its radius is "The volume of a sphere is proportional to R^3." So if you double the radius of a sphere the sphere's volume increases not by a factor of 2 but by a factor of 8 (since $2^3 = 8$).

So although proportionality relationships involving quantities raised to a power require some extra vigilance, forming ratios of such relationships has the same benefits (canceling of constants) discussed above. To see how that works, form the ratio of the proportionality relationships for the areas of two circles:

$$\frac{A_2 \propto R_2^2}{A_1 \propto R_1^2}.$$

Now turn the proportionality relationships into equations by explicitly writing the constant of proportionality, which we'll call "w" (remember, when you form ratios of proportionality relationships, you can call the constants whatever you like):

$$\frac{A_2}{A_1} = \frac{w R_2^2}{w R_1^2} = \frac{\cancel{w} R_2^2}{\cancel{w} R_1^2} = \left(\frac{R_2}{R_1}\right)^2.$$

So the ratio of proportionality relations gives exactly the same result as that obtained by using the ratio of equations.

Example: Compare the volumes of two spheres, where one has five times the radius of the other.

From Eqs. 1.6, you know that $V \propto R^3$. Writing this relationship as an equation with the proportionality constant "s," and comparing two different spheres (say, sphere 1 and 2) by dividing equations gives

$$\frac{V_2}{V_1} = \frac{s R_2^3}{s R_1^3} = \frac{\cancel{s} R_2^3}{\cancel{s} R_1^3} = \left(\frac{R_2}{R_1}\right)^3.$$

Since you know that one of the spheres (say, sphere 2) has a radius five times the other, you can use the mathematical substitution $R_2 = 5R_1$. Plugging $5R_1$ in for R_2 in the expression above and simplifying gives

$$\frac{V_2}{V_1} = \left(\frac{5\cancel{R_1}}{\cancel{R_1}}\right)^3 = \left(\frac{5}{1}\right)^3 = 5^3 = 125.$$

That is, a sphere with five times larger radius has 125 times larger volume.

Exercise 1.11. **Hubble's Law states that a galaxy's velocity (v) and its distance from us (d) are directly proportional: $v \propto d$. If galaxy Z is 50 Mpc (megaparsecs) away from us and galaxy Y is 800 Mpc away, how do their velocities compare?**

1.2.6 Inverse proportionality relationships

An *inverse* proportionality means the relationship is reversed: as one quantity gets larger, the other gets smaller. For example, the wavelength (λ) and frequency (f) of light are inversely proportional. This is represented mathematically as $\lambda \propto 1/f$, or equivalently, $f \propto 1/\lambda$. This relationship will be explored in more detail in Section 3.1; here's an example of how to use such a relationship.

Example: Wavelength and frequency of light are inversely proportional. Red visible light has a wavelength 75% larger than blue visible light. How do their frequencies compare?

Since λ and f are inversely proportional, you can predict the answer qualitatively: since red light has a *larger* wavelength, it must have a *smaller* frequency – this is the essence of an inverse relationship. To be more quantitative, you can write the inverse relationship as an equation:

$$f \propto \frac{1}{\lambda}, \text{ or } f = c \times \frac{1}{\lambda} = \frac{c}{\lambda}.$$

Now compare red and blue by dividing their equations:

$$\frac{f_{red}}{f_{blue}} = \frac{\cancel{c}/\lambda_{red}}{\cancel{c}/\lambda_{blue}} = \frac{1/\lambda_{red}}{1/\lambda_{blue}} = \frac{1}{\lambda_{red}} \times \frac{\lambda_{blue}}{1} = \frac{\lambda_{blue}}{\lambda_{red}}.$$

Next, translate the information given in the problem into a mathematical relationship: "The wavelength of red light is 75% *more* than the wavelength of blue light." That means λ_{red} is 75% *more than 100%* of λ_{blue} (because 100% would mean they are the same). Thus the relationship is $\lambda_{red} = 175\% \times \lambda_{blue}$, or $\lambda_{red} = 1.75\lambda_{blue}$. Plugging this substitution into the expression above,

$$\frac{f_{red}}{f_{blue}} = \frac{\cancel{\lambda_{blue}}}{1.75\cancel{\lambda_{blue}}} = \frac{1}{1.75} = 0.571,$$

so the frequency of red light is about 57% that of blue light. This agrees with the prediction that red light must have a smaller frequency.

Several other instances of inverse proportionality appear in this book. One example is that the force of gravity between two objects is inversely proportional to the square of the distance between the centers of those objects:

$$F_g \propto \frac{1}{R^2},$$ (1.8)

in which F_g represents the force of gravity and R represents the distance between the centers of the objects.

Example: How does the force of gravity between two objects change if the distance between the objects doubles?

Writing the proportionality relationship of Eq. 1.8 as an equation with proportionality constant "z" for both the far and the near distance gives:

$$F_{g,\,far} = z \times \frac{1}{R_{far}^2} = \frac{z}{R_{far}^2},$$

$$F_{g,near} = z \times \frac{1}{R_{near}^2} = \frac{z}{R_{near}^2},$$

which can be compared by dividing:

$$\frac{F_{g,far}}{F_{g,near}} = \frac{\frac{1}{R_{far}^2}}{\frac{1}{R_{near}^2}} = \frac{1}{R_{far}^2} \cdot \frac{R_{near}^2}{1} = \frac{R_{near}^2}{R_{far}^2} = \left(\frac{R_{near}}{R_{far}}\right)^2.$$

Since the objects doubled in distance from one another, you can write $R_{far} = 2R_{near}$. Making this substitution in the expression above gives

$$\frac{F_{g,far}}{F_{g,near}} = \left(\frac{R_{near}}{2R_{near}}\right)^2 = \left(\frac{1}{2}\right)^2 = \frac{1}{4},$$

which means that if you double the distance between two objects, the gravitational force between them drops to one-quarter of its previous strength. This kind of mathematical relationship, where a quantity depends on the inverse of the distance squared, is referred to as an "inverse-square law" and is common in physics. In this book, you'll find such relationships in Sections 2.1 and 5.2.

Exercise 1.12. Ultraviolet light has a frequency that is about one hundred times that of infrared light. Which has a larger wavelength, and by how many times?

Exercise 1.13. If one moon orbits at a certain distance from its planet, and another moon orbits three times farther from the same planet, compare the planet's force of gravity on the closer moon to the planet's force of gravity on the farther moon.

1.3 Rate problems

In many science classes, you may be asked to calculate things such as how long it takes to travel a certain distance at a fixed speed, or how long a star will live if it burns its fuel at a certain rate. These "rate" problems are particularly prevalent in astronomy, and a fixed speed that frequently comes up is c, the speed of light.

1.3.1 Distance, speed, and time problems

You probably already have an intuitive sense for how to do distance, speed, and time problems. For example, how far will you go if you ride your bicycle at 10 km/hr for 2 hours? Well, each hour you cover 10 km, so you will go a total distance of 20 km. Or, how long will it take you to walk 12 km at a speed of 3 km/hr? Since you will cover 3 km each hour, it will take you 4 hours to go 12 km. You may have done those problems in your head, but what exactly did you *do* in your head to get the answers?

Most likely, you intuitively used a general relationship between distance, speed, and time that can be written like this:

$$\text{distance} = \text{speed} \times \text{time}. \tag{1.9}$$

Here are detailed descriptions of each of the terms and operations in the distance equation:

distance   The total distance covered during the time the object is moving, with dimensions of length. The standard (SI) units of length are meters; other units often used in astronomy are kilometers (km), astronomical units (AU), parsecs (pc), and light years (ly).

speed   The rate at which the object is moving,[5] with dimensions of length over time. The SI units of speed are meters per second (remember that "per" means "divided by"); other popular units include miles per hour and kilometers per hour. In Eq. 1.9, speed is assumed to remain constant.

x   Multiplication of these two quantities makes sense because distance increases directly with both speed and time. It also ensures that the units work out properly; for example (km/s) × s = km.

time   The total time during which the object is moving. The SI units of time are seconds; other units include minutes, hours, and years.

---

[5] In some astronomy texts, "speed" is also called "velocity," although velocity is actually a vector that includes both speed and direction.

Using this relationship, you can determine the value of any one of these parameters as long as you're given the other two. By applying this equation to the walking and bicycling examples, you can see that its results match with the intuitive answers already given.

For the case of bicycling at 10 km/hr for 2 hours, the speed and time have been provided, and you are asked to calculate the distance. The equation already has distance alone on the left side (that is, the equation is "solved for distance"), so you can just plug in the speed and time:

$$\text{distance} = \text{speed} \times \text{time} = \left(10 \, \frac{\text{km}}{\text{hr}} \right) \times 2 \, \text{hr} = 20 \text{ km}.$$

For the second example, walking 12 km at 3 km/hr, the distance and speed have been given, and you are asked to calculate the time. Time is not already by itself on one side of the equation, so you can proceed in one of two ways. The instinct of many students is to plug in the values first, and then solve for time. Although we don't recommend this approach, if you solve carefully and carry your units through, this approach will give you the correct answer, as follows:

$$12 \text{ km} = \left(3 \, \frac{\text{km}}{\text{hr}} \right) \times \text{time}.$$

Now, dividing both sides by 3 km/hr to get time by itself,

$$\frac{12 \text{ km}}{3 \, \frac{\text{km}}{\text{hr}}} = \frac{3 \, \frac{\text{km}}{\text{hr}}}{3 \, \frac{\text{km}}{\text{hr}}} \times \text{time} = \text{time}, \tag{1.10}$$

and simplifying the numbers and units separately gives

$$\text{time} = \frac{12 \text{ km}}{3 \, \frac{\text{km}}{\text{hr}}} = \left(\frac{12}{3} \right) \left(\frac{\text{km}}{\frac{\text{km}}{\text{hr}}} \right) = 4 \left(\text{km} \, \frac{\text{hr}}{\text{km}} \right) = 4 \text{ hr}.$$

Notice that to simplify a compound fraction (that is, a fraction within a fraction such as $\frac{\text{km}}{\text{km/hr}}$), you can invert the denominator and multiply it by the numerator. This "invert and multiply" simplification can be used with numbers and units alike, so for example $\frac{4}{2/3} = 4(\frac{3}{2}) = \frac{12}{2} = 6$.

The alternative way to do this problem, which is quicker and leaves fewer opportunities for mistakes, is to solve Eq. 1.9 for the desired quantity *first*, before plugging in values. In this case, time is the desired quantity. You can "solve for time" by dividing both sides of the Eq. 1.9 by speed:

$$\frac{distance}{speed} = \frac{\cancel{speed} \times time}{\cancel{speed}} = time,$$

$$time = \frac{distance}{speed}. \tag{1.11}$$

Now that you have time by itself on the left side of the equation, you're ready to plug in your numerical values. Remember that plugging in numbers should always be the very last step:

$$time = \frac{12 \text{ km}}{3 \frac{km}{hr}} = \left(\frac{12}{3}\right)\left(\frac{km}{\frac{km}{hr}}\right) = 4 \left(\cancel{km}\frac{hr}{\cancel{km}}\right) = 4 \text{ hr.}$$

One reason that you're far better off solving for the quantity you're seeking *before* plugging in numerical values can be seen by comparing Eq. 1.10 (plugging in before solving) to Eq. 1.11 (solving before plugging in). You can use Eq. 1.11 to solve *any* problem in which you're given the distance and speed and asked to find the time, so this equation is the solution to a multitude of problems. But if you plug in first, as in Eq 1.10, you have the solution to *only one* particular problem.

It's also very important for you to realize that time and distance units must be consistent throughout Eq. 1.9 (and Eq. 1.11) in order to do the calculation with numerical values. That is, the units in the time term must match the time units in the denominator of the speed units, and the units of the distance term must match the units of length in the numerator of the speed units. If they do not, then you will have to perform a unit conversion to make them match before plugging in, or as part of your calculation. For instance, if you have a distance in parsecs and a speed in kilometers per second, you will have to convert parsecs to kilometers or kilometers to parsecs in order to plug in the values. Similarly, if you have a speed in meters per second and a time in years, you will need to convert seconds to years or years to seconds. You can convert the units before you plug in the values, or you can plug in the mismatched units and then include a conversion factor as part of the calculation.

Example: How far does light (which travels at a speed of 3×10^8 m/s) travel in 1 year?

To find the distance, you'll have to multiply a speed (given in units of meters per second) by a time (given in units of years). This requires converting either seconds to years in the speed term or years to seconds in the time term before you can multiply. To do this latter conversion, you can use the conversion factor between seconds and years, 1 year ↔ 31.5 million seconds, derived in Section 1.1. This unit conversion can be done as a separate step before plugging the values of speed and time into the equation, or the conversion factor can be included right in the problem like this:

$$\text{distance} = \text{speed} \times \text{time} = \left(3 \times 10^8 \ \frac{m}{s}\right) \times 1 \ \cancel{yr} \times \frac{31,500,000 \ \cancel{s}}{1 \ \cancel{yr}}.$$

Notice the seconds cancel even though the terms are not adjacent (since multiplication is commutative, the order of the terms does not matter). Writing the numbers in scientific notation,[6] you get

$$\text{distance} = (3 \times 10^8) \times (3.15 \times 10^7) \ m = 9.5 \times 10^{15} \ m.$$

Thus you have calculated that light travels about 10 quadrillion meters – or 10 trillion kilometers – in one year. This is the definition of 1 light year, so in this example you have derived another useful conversion factor: 9.5×10^{15} m ↔ 1 ly.

Exercise 1.14. Calculate the time it takes for a train traveling at 100 km/hr to go 70 miles.

Exercise 1.15. Earth moves in its orbit around the Sun at a speed of 29.8 km/s. How many meters does our planet move in one minute?

1.3.2 Amount, rate, and time problems

The rate equation you have been using, distance = speed × time, can be generalized to a relationship that is useful in many other circumstances:

$$\text{amount} = \text{rate} \times \text{time}. \tag{1.12}$$

In this generalized equation, "rate" doesn't just refer to speed in the sense of distance divided by time; it can also refer to other kinds of speed such as how many pages you can read per hour or how much money you spend per day. And you probably have an intuitive sense for solving such generalized rate problems in your everyday life. For example, if you were asked to calculate how many biscuits you'd consume if you ate them at a rate of 2 biscuits per day for a week, you might well recognize – without having to write down any math – that you'd consume 14 biscuits. But what you did in your head to arrive at that answer is completely analogous to the steps we applied to the distance, rate, and time problems. In this example you are given the time (7 days) and the

---

[6] Scientific notation is discussed in Section 1.4.

rate (2 biscuits per day), and you are asked to calculate the amount. Plugging directly into Eq. 1.12 gives

$$\text{amount} = \text{rate} \times \text{time} = 2 \; \frac{\text{biscuits}}{\text{day}} \times 7 \; \text{days} = 14 \; \text{biscuits}.$$

Here's how you can apply Eq. 1.12 to an example that does not lend itself so readily to computation in your head.

Example: If the Sun has 9×10^{28} kg of hydrogen available as fuel, and if it uses up that fuel at a rate of 6×10^{11} kg/s, how long will it take the Sun to use up all of its available fuel?

You are given the amount of fuel and rate of fuel consumption, and asked to calculate the time. Employing the method of first rearranging Eq. 1.12 to isolate time on one side, analogous to Eq. 1.11, you should get

$$\frac{\text{amount}}{\text{rate}} = \frac{\text{rate} \times \text{time}}{\text{rate}} = \text{time}$$

or

$$\text{time} = \frac{\text{amount}}{\text{rate}}. \tag{1.13}$$

Now plugging in values only after the rearranging is done, grouping numbers and units, and simplifying gives

$$\text{time} = \frac{9 \times 10^{28} \; \text{kg}}{6 \times 10^{11} \; \frac{\text{kg}}{\text{s}}} = \left(\frac{9}{6}\right)\left(\frac{10^{28}}{10^{11}}\right)\left(\text{kg}\frac{\text{s}}{\text{kg}}\right) = 1.5 \times 10^{17} \; \text{s}.$$

This is the remaining lifetime of the Sun, after which the Earth will become uninhabitable. Should you worry about whether this will happen this week? In your lifetime? In your great-grandchildren's lifetime? In units of seconds, the Sun's lifetime is such a large number that most people don't have a sense for how much time it is. In the problems at the end of this chapter, you will have a chance to convert this time to the more useful unit of years.

In this section, you have seen that you can use the rate relationships, Eqs. 1.9 and 1.12, to calculate any of those quantities if you know the other two. You may need to rearrange the equation to isolate the quantity you are trying to calculate, and you may also need to incorporate a unit conversion to get a consistent set of units for canceling. In the next example, you'll see how to combine all these techniques in a single problem.

Example: Imagine that you wish to count each of the 300 billion or so stars in our galaxy within one (long) human lifetime of 90 years. How fast would you have to count? That is, what counting rate (in units of stars per second) would allow you to count 300 billion stars in 90 years?

As in all problems, a very good way to begin is to write down exactly what you're given, what you're trying to find, and what relationship exists between those quantities. In this case, you're given the amount to count (300 billion stars) and the time (90 years), and you are asked to calculate the rate. You also know that the generalized rate equation, Eq. 1.12, relates amount, time, and rate. So you should start by solving that equation for rate:

$$\frac{\text{amount}}{\text{time}} = \frac{\text{rate} \times \cancel{\text{time}}}{\cancel{\text{time}}} = \text{rate}.$$

This is in fact the general definition of a rate: an amount per time. Now plugging in the values you were given, and simultaneously introducing the conversion factor for changing years into seconds to obtain the desired units gives

$$\text{rate} = \frac{300 \times 10^9 \text{ stars}}{90 \cancel{\text{years}}} \times \frac{1 \cancel{\text{year}}}{3 \times 10^7 \text{ s}} = \left(\frac{300 \times 10^9}{3 \times 90 \times 10^7} \right) \left(\frac{\text{stars}}{\text{s}} \right)$$

in which we've rounded the number of seconds in one year to 30 million for simplicity. Grouping numbers and simplifying, this is

$$\text{rate} = \left(\frac{300}{270} \right) \left(\frac{10^9}{10^7} \right) \left(\frac{\text{stars}}{\text{s}} \right) = 1.1 \times 10^2 \frac{\text{stars}}{\text{s}} = 110 \text{ stars/s}.$$

That is, you'd have to count over 100 stars each second for almost a century, with no breaks for eating or sleeping, to count the stars just in our Milky Way galaxy. And if that number of stars seems unfathomable, remember that ours is only one of the hundreds of billions of galaxies in the known Universe.

Exercise 1.16. How many pages per hour must you read in order to finish a book with 217 pages in 8 hours? Convert your answer into units of pages per minute.

Exercise 1.17. At the rate you calculated in the previous exercise, how long would it take you to read a 350-page book?

1.4 Scientific notation

It is an inescapable consequence of the immense scale of the Universe that astronomy deals with huge numbers. Our Sun has a mass of approximately 2,000,000,000,000,000,000,000,000,000,000 kilograms, there are about 300,000,000,000 stars in our Milky Way galaxy, and there are between 50,000,000,000 and 1,000,000,000,000 galaxies in the observable Universe. You can express huge numbers such as these using words, such as two thousand billion billion billion kilograms, and three hundred billion stars, but this is still unwieldy. And you can't do calculations with numbers that are written out in words. The most succinct and flexible way to write and manipulate very large (and very small) numbers is to use scientific notation.

Of course, the subject of scientific notation is covered fairly early in the curriculum of most schools, so you may be entirely comfortable with numbers expressed as 2×10^{30} or 6.67×10^{-11}. If so, feel free to skip this section. But if it's been a few years since you've encountered scientific notation, or if you have any doubt at all about the difference between 6×10^{-3} and -6×10^3, or how to calculate $(8 \times 10^7)/(2 \times 10^{12})$ in your head, this section may be helpful for you.

1.4.1 Coefficient, base, and exponent

In scientific notation, the very large number 300,000,000 (which is the number of meters light travels in one second) is written as 3×10^8, and the very small number 0.0000000000667 (which is the Universal Gravitational Constant in standard units) is written as 6.67×10^{-11}. As shown in Figure 1.3, each of these expressions consists of three numbers called the coefficient, the base, and the exponent. The standard base for scientific notation is 10. Exponents are usually integers and can be positive or negative. The coefficient can be any number at all. If you see a number in scientific notation in which the coefficient is missing, such as 10^6, it is important to remember that a coefficient of 1.0 is implicit. That is, $10^6 = 1 \times 10^6$.

Many astronomy texts use "normalized" scientific notation in which the decimal point in the coefficient always appears immediately to the right of the leftmost non-zero digit. So although 3.5×10^4 and $35. \times 10^3$ represent exactly the same number, astronomy texts are more likely to use the first version of this number. In normalized scientific notation, the coefficient is always between one and ten, and the exponent is called the "order of magnitude" of the number.

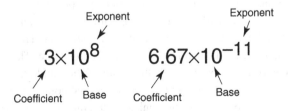

Figure 1.3 The elements of a number in scientific notation.

If you think about the mathematical operations represented in scientific notation, you can understand why these numbers are written this way. First, consider the number 300,000,000 or 3×10^8, and recall that 10^8 is simply $10 \times 10 \times 10 \times 10 \times 10 \times 10 \times 10 \times 10$, which is 100,000,000. So 3×10^8 is just 3 times 100,000,000, which is 300,000,000.

The same logic applies to the number 0.0000000000667 or 6.67×10^{-11}, but in this case the number 10^{-11} is the very small number $\frac{1}{10^{11}}$, or $\frac{1}{100,000,000,000}$, or 0.00000000001. So 6.67×10^{-11} means 6.67 times 0.00000000001, which is 0.0000000000667.

One thing to remember when you're dealing with numbers written in scientific notation is that a negative sign in front of the coefficient (such as -6×10^3) means that the number is negative, but a negative sign in the exponent (such as 6×10^{-3}) does not have any effect on whether the number is positive or negative. So what does a negative exponent mean? Simply this: the more negative the exponent, the closer the value of the number is to zero. So 6×10^{-3} is a small number, and 6.3×10^{-11} is a very very small number. In astronomy, you are unlikely to encounter many negative numbers, but you are very likely to see negative exponents. For example, the values of some of the physical constants, wavelengths of light, and masses of atoms are all very small and are often written using scientific notation with negative exponents.

Example: Identify the base, coefficient, and exponent in the numbers (a) 150×10^6 and (b) 1.6×10^{-19}.

The coefficient is the number in front including any negative signs (though in this case, both numbers are positive), so the coefficients for (a) and (b) are 150 and 1.6, respectively. The base is 10 for both, the standard for scientific notation. The exponent is the power that 10 is raised to, including any negative signs, so the exponents are 6 and -19, respectively.

Note that the number in (a) above is not in normalized scientific notation because the coefficient (150) is not between 1 and 10. Sometimes you may wish to move the decimal point in the coefficient – perhaps to put it in normalized notation, or to facilitate comparing with other numbers in scientific notation, or to allow you to do a calculation in your head. This comes up a lot, so it's a good idea to be comfortable with this procedure. The key to keep in mind is that you are not changing the *value* of the number; you are only changing the way it looks. So if you move the decimal point in the coefficient, the value of the number will change unless you adjust the exponent to compensate. For example, if you move the decimal point in the coefficient to the *left* some number of places, then you are making the coefficient *smaller* by that many powers of 10, so you must *increase* the exponent by the same number of powers of 10 to compensate. This ensures that the overall value of the entire number is not changed.

Example: Express 150×10^6 in normalized scientific notation.

The first step is to change "150." (the decimal point after the zero was implicit) to "1.50," which requires moving the decimal point two places to the left. This decreases the value of the coefficient by a factor of 100 (two powers of 10). You must then increase the value of the rest of the number (10^6) by two powers of 10, to 10^8. Thus the remodeled number is 1.5×10^8, which has exactly the same value as 150×10^6.

On the other hand, if you move the decimal point in the coefficient to the *right* some number of places, you are making the coefficient *larger* by that many powers of 10, so you must *decrease* the exponent by the same number of powers to compensate. For example, if you have the number 0.026×10^3, you can put this into normalized scientific notation by moving the decimal point of the coefficient two places to the right (so 0.026 becomes 2.6). This is equivalent to multiplying the coefficient by a factor of 100 (two powers of 10), so to compensate you need to reduce the rest of the number by a factor of 100. To do that, you can reduce the exponent by two, which turns 10^3 into 10^1. Thus $0.026 \times 10^3 = 2.6 \times 10^1$.

Here are some equivalent expressions (not in normalized notation) for the numbers used in the preceding examples:

$$150 \times 10^6 = (150 \times 10) \times 10^{(6-1)} = 1500 \times 10^5$$

and

$$1.6 \times 10^{-19} = (1.6 \times 1000) \times 10^{(-19-3)} = 1600 \times 10^{-22}.$$

Exercise 1.18. Write the following numbers in scientific notation with the coefficients and exponents given.

(a) **Coefficient = 6022; exponent = 20.**
(b) **Coefficient = 0.91; exponent = −6.**
(c) **Express each of the numbers above in *normalized* scientific notation.**

1.4.2 Converting numbers in scientific notation

Converting numbers to and from scientific notation is straightforward as long as you pay careful attention to which direction you're moving the decimal point. Here are the rules for converting numbers from scientific notation (such as 3×10^8) into decimal notation (such as 300,000,000). The following procedure applies to base-10 numbers, and "first" means "farthest to the left" while "last" means "farthest to the right":

- Write down the coefficient without the base or the exponent.
- If no decimal point is shown for the coefficient, insert a decimal point at the end (that is, to the right) of the last digit.
- If the exponent is positive, move the decimal point to the right (inserting zeros as needed) by the number of places indicated by the exponent.
- If the exponent is negative, move the decimal point to the left (inserting zeros as needed) by the number of places indicated by the exponent.

Example: Express the numbers 3×10^8 and 6.67×10^{-11} in decimal notation.

In the case of 3×10^8, first write down the coefficient (3), add a decimal point at the end (3.). Then move that decimal point 8 places to the right, because the exponent is 8. This gives 300,000,000 (note the decimal point can be omitted now), which is the value of the number 3×10^8 in decimal notation.

For the number 6.67×10^{-11}, begin by writing down the coefficient (6.67); in this case the decimal point is already there, so there's no need to add it. Since the exponent is negative, you then move the decimal point 11 places to the left, which gives 0.0000000000667 (much less than one).

As you might expect, the process for converting numbers from decimal notation into scientific notation is just the reverse (as before, this applies to base-10 numbers only):

- Write down the decimal number you wish to convert to scientific notation. If no decimal point is shown in your number, insert a decimal point at the end (that is, to the right) of the last digit.
- Below your number, write a "new" version of the number. For this version, write the first non-zero digit of your number, followed by a decimal point, followed by all the other digits (if any) of your number. This will be the coefficient of your number in scientific notation.
- Count the number of places you would have to move the decimal point (either to the right or to the left) to turn the new version of your number into the original version.

- If you had to move the decimal point to the *right* to arrive at the position shown in the original version of your number, your exponent is a *positive* number equal to the number of places the decimal point has moved.

- If you had to move the decimal point to the *left* to arrive at the position shown in the original version of your number, your exponent is a *negative* number equal to the number of places the decimal place has moved.

- To the right of the coefficient you've just written, write $\times 10^0$ and put your exponent in place of the parentheses.

Example: Express 412,000 in scientific notation.

Begin by writing down the original version of your number with its original decimal place (412,000.), and below it the new version with only one non-zero digit to the left of the decimal point (4.12000). To turn this new version of your number into the original version, you would have to move the decimal point five places to the right, so your exponent is 5, and your number in scientific notation is 4.12000×10^5. Unless you're keeping track of significant figures, you can leave off the trailing zeroes and write this as 4.12×10^5.

As a check on the sign of your exponent, ask yourself if the value of your number is very big (either large positive or large negative) or very small (i.e. close to zero). If it's a big number (i.e. far from zero) the exponent is positive; for a number very close to zero, the exponent is negative.

After completing either type of conversion, it's a good idea to use the reverse conversion process as a check to make sure that you get back to your original number. Indeed, starting with 4.12×10^5 and moving the decimal point five places to the right while dropping the "$\times 10^5$" yields 412,000.

Exercise 1.19. Express the following numbers in scientific notation:
(a) 3,300 (b) −3,300 (c) 100,000,000,000 (d) 0.0000000048 (e) −0.0000000048.

Exercise 1.20. Write the following numbers out in decimal notation:
(a) 9.3×10^7 (b) $−9.3 \times 10^7$ (c) 9.3×10^{-7} (d) 1×10^7 (e) 10×10^7 (f) 10^7 (g) 5.2×10^0.

1.4.3 Numbers as words

If you were to read in an astronomy book that there are three hundred billion stars in our Milky Way galaxy, how can you make sense of that number? Certainly, if you need to do any calculations with numbers given in this format, you must be able to translate between words, decimal notation, and scientific

notation. Here's a table of some of the words describing large numbers that
come up in astronomy:

Words	Decimal notation	Scientific notation
Thousand	1,000	10^3
Million	1,000,000	10^6
Billion	1,000,000,000	10^9
Trillion	1,000,000,000,000	10^{12}
Quadrillion	1,000,000,000,000,000	10^{15}

Example: Write three hundred billion in scientific and decimal notation.

Three hundred billion can be thought of as three hundred *times* one billion, or
300×10^9. (In normalized scientific notation, this is 3×10^{11}.) Written out in
decimal notation, this is 300,000,000,000. A trick for going the other direction,
from decimal notation to words, is to start at the left of the number and read to
the right in groupings of three zeroes at a time. In this case, first you see "300"
("three hundred"), followed by ",000" (making it "three hundred thousand"),
followed by another ",000' (making it "three hundred million"), then one final
",000" (making it "three hundred billion").

**Exercise 1.21. Complete the following table to express each quantity all
three ways.**

Words	Decimal notation	Scientific notation
Three million	(a)	(b)
(c)	12,000,000,000,000	(d)
(e)	(f)	1×10^5 (or just 10^5)
Half a billion	(g)	(h)
(i)	95	(j)

1.4.4 Calculations with scientific notation

One of the advantages of using scientific notation is the ease with which you
can multiply and divide numbers without using a calculator. To do such calculations easily, treat the coefficients and exponents separately as shown in the
following examples.

Example: What is $(2 \times 10^4) \times (4 \times 10^3)$?

To multiply two numbers expressed in scientific notation, you simply *multiply* the coefficients and *add* the exponents. So if you want to multiply 2×10^4 by 4×10^3, you can just multiply the coefficients ($2 \times 4 = 8$) and add the exponents ($4 + 3 = 7$) to get the correct result of 8×10^7. To see why this works, you can write the numbers out in decimal notation:

$$(2 \times 10^4) \times (4 \times 10^3) = (20{,}000) \times (4{,}000) = 80{,}000{,}000 = 8 \times 10^7.$$

Notice that when you multiplied 20,000 by 4,000, the result had to have all the zeroes of each factor – that is, the *sum* of the number of zeroes, which is the sum of the exponents. That is why you add exponents when you multiply quantities in scientific notation.

When one of the exponents is negative, the process works in the same way. You still add exponents, taking care to retain the negative sign.

Example: What is $(3 \times 10^3) \times (2 \times 10^{-10})$?

$$(3 \times 10^3) \times (2 \times 10^{-10}) = (3 \times 2) \times (10^{(3+(-10))}) = 6 \times 10^{3-10} = 6 \times 10^{-7}.$$

A similar approach works for dividing two numbers expressed in scientific notation, except that in this case you *divide* the coefficients and *subtract* the exponents.

Example: What is $(2 \times 10^5) \div (4 \times 10^3)$?

Divide the coefficients ($2 \div 4 = 0.5$) and subtract the exponents ($5 - 3 = 2$) to get the correct result of 0.5×10^2. Writing it out in decimal notation shows why this works:

$$\frac{2 \times 10^5}{4 \times 10^3} = \frac{200{,}000}{4{,}000} = \frac{200{,}\cancel{000}}{4{,}\cancel{000}} = \frac{200}{4} = \frac{2 \times 10^2}{4} = \left(\frac{2}{4}\right) \times 10^2$$
$$= 0.5 \times 10^2.$$

To express this answer in normalized scientific notation, move the decimal point of the coefficient one place to the right (making it ten times bigger), and hence adjust the exponent down by one power (making it ten times smaller): $0.5 \times 10^2 = (0.5 \times 10) \times (10^{2-1}) = 5.0 \times 10^1$.

When one of the exponents is negative, dividing works just the same. You still subtract exponents, taking care to retain the negative sign.

Example: What is $(2 \times 10^3) \div (3 \times 10^{-10})$?

$$\frac{2 \times 10^3}{3 \times 10^{-10}} = \frac{2}{3} \times \frac{10^3}{10^{-10}} = \frac{2}{3} \times 10^{3-(-10)} = \frac{2}{3} \times 10^{(3+10)} = 0.67 \times 10^{13}.$$

Again, for normalized scientific notation, you can move the decimal point of the coefficient one place to the right, remembering to adjust the exponent down by one power to compensate: $0.67 \times 10^{13} = (0.67 \times 10) \times (10^{13-1}) = 6.7 \times 10^{12}$.

Here's a shortcut that frequently comes in handy when dividing numbers in scientific notation: you can move a base and its exponent across the fraction bar, as long as you flip the sign of the exponent. Here are several examples illustrating this procedure:

$$\frac{1}{10^3} = \frac{10^{-3}}{1} = 10^{-3}, \qquad \frac{1}{10^{-6}} = 10^6, \qquad \text{and} \qquad 10^{-7} = \frac{1}{10^7}$$

This is true because changing the *sign* of the exponent (e.g. 10^3 to 10^{-3}) and inverting a fraction mean exactly the same thing: taking a mathematical inverse (also called the reciprocal). So if you change the sign of the exponent *and* invert the fraction, these changes nullify each other, and there is no net change to the underlying quantity.

You can also see why this works by the division rule shown above:

$$\frac{1}{10^3} = \frac{10^0}{10^3} = 10^{0-3} = 10^{-3}.$$

Note that you cannot move a coefficient unscathed across the fraction bar. That is, 5×10^{-3} is *not* equal to $\frac{1}{5 \times 10^3}$. You can either leave the coefficient where it is, or you can also take the inverse of the coefficient if it crosses the fraction bar, so $5 \times 10^{-3} = \frac{5}{10^3} = \frac{1}{\frac{1}{5} \times 10^3}$. Since the number 1 is its own inverse, it alone can cross the fraction bar unchanged: $1 \times 10^{-17} = \frac{1}{1 \times 10^{17}}$. But since you can omit the coefficient of 1, this could be written simply $10^{-17} = \frac{1}{10^{17}}$.

Exercise 1.22. Perform the following operations without using a calculator. Express your answers in normalized scientific notation:
(a) $(3 \times 10^5) \times (10^4)$ (b) $(6 \times 10^{-6}) \times (3 \times 10^4)$
(c) $(6 \times 10^{-6}) \div (3 \times 10^4)$ (d) $(6 \times 10^6) \div (10^4)$

1.4.5 Order-of-magnitude estimation

If you're trying to multiply or divide numbers with coefficients that are not integers, doing this process in your head is not always easy. In such cases, you can get an approximate answer by rounding the coefficients to the nearest integer before multiplying or dividing. And for a "rough order of magnitude" estimate (sometimes called a ROM), you may even consider rounding any coefficient less than about 3 (roughly the square root of 10) down to 1 and any coefficient greater than 3 up to 10 to make the multiplication or division even easier.

Example: Use rounding to estimate $(1.087 \times 10^{21}) \times (5.5 \times 10^3)$.

Rounding to the nearest integers makes this $(1 \times 10^{21}) \times (6 \times 10^3)$, which gives a result of 6×10^{24}, very close to the calculator's answer of 5.9785×10^{24}. Even using the order-of-magnitude approach gives $(1 \times 10^{21}) \times (10 \times 10^3) = 10 \times 10^{24}$, which is a usable ROM for a quick estimate.

Exercise 1.23. Use both techniques of integer rounding and ROM estimation to perform the following calculations without a calculator.
(a) $(1.23 \times 10^5) \times (4.56 \times 10^4)$ (b) $(9.87 \times 10^{-6}) \div (6.54 \times 10^4)$
(c) $(6.6 \times 10^{-6}) \times (2.2 \times 10^4) \div (1.8 \times 10^4)^2$

1.4.6 Raising numbers to powers

Many of the equations you will encounter in astronomy will have a power in them – usually a square, cube, or fourth power. For example, c (the speed of light) is squared in $E = mc^2$, R (radius of a sphere) is cubed in $V = \frac{4}{3}\pi R^3$, and T (temperature of a body emitting thermal radiation) is raised to the fourth power in $L = 4\pi R^2 \sigma T^4$. If you plug in a very large or very small numerical value for one of these terms that is raised to a power, you will need to know how to raise a number in scientific notation to a power. The key is to apply the power to the coefficient *and* the exponent separately (and in different ways): Raise the coefficient to that power and *multiply* the exponent by the power. You can see why this works in the following examples.

Example: What is $(7 \times 10^5)^2$?

$$(7 \times 10^5)^2 = 700{,}000 \times 700{,}000 = 490{,}000{,}000{,}000 = 49 \times 10^{10},$$

which is equivalent to

$$(7 \times 10^5)^2 = 7^2 \times (10^5)^2 = 49 \times 10^{(5 \times 2)} = 49 \times 10^{10}.$$

This same process also works with negative exponents:

Example: What is $(2 \times 10^{-4})^3$?

$$(2 \times 10^{-4})^3 = 0.0002 \times 0.0002 \times 0.0002 = 0.000000000008 = 8 \times 10^{-12},$$

which is equivalent to

$$(2 \times 10^{-4})^3 = 2^3 \times (10^{-4})^3 = 8 \times 10^{(-4 \times 3)} = 8 \times 10^{-12}.$$

When you rearrange an equation and solve for a variable, you sometimes end up taking a square root or a cube root, so you need to know how to take a root of a number in scientific notation. It helps to realize that the root applies to both the coefficient and the exponent, so $\sqrt{6 \times 10^{12}} = \sqrt{6} \times \sqrt{10^{12}}$. A root of a quantity is equivalent to raising the quantity to a fractional power – the exponent becomes the *inverse* of the degree of the root. For example, taking a square root of a number is the same as raising that number to the power $\frac{1}{2}$ (so $\sqrt{x} = x^{\frac{1}{2}}$). Likewise, taking the cube root of a number is the same as raising that number to the power $\frac{1}{3}$ (so $\sqrt[3]{x} = x^{\frac{1}{3}}$).

Example: What is $\sqrt[2]{9 \times 10^4}$?

$$\sqrt[2]{9 \times 10^4} = \sqrt[2]{9} \times \sqrt[2]{10^4} = 3 \times (10^4)^{\frac{1}{2}} = 3 \times 10^{(4 \times \frac{1}{2})} = 3 \times 10^2.$$

Example: What is $\sqrt[3]{8 \times 10^9}$?

$$\sqrt[3]{8 \times 10^9} = \sqrt[3]{8} \times \sqrt[3]{10^9} = 2 \times (10^9)^{\frac{1}{3}} = 2 \times 10^{(9 \times \frac{1}{3})} = 2 \times 10^3.$$

You may be able to do these in your head if the numbers are whole multiples of the root, such as the two examples above. Otherwise you'll need a calculator.

Exercise 1.24. Perform the following operations without a calculator:
(a) $(2 \times 10^{-3})^4$ (b) $(4 \times 10^{14})^{-\frac{1}{2}}$ (c) $\sqrt[3]{1 \times 10^{-15}}$

1.4.7 Calculator issues

If you require precise answers and you're allowed to use a calculator, be sure to enter numbers expressed in scientific notation into your calculator properly. Entering a number such as 6×10^{24} into your calculator by pressing "6" and then pressing "×" and then "10" followed by "^" and "24" is asking for trouble, because if this number is part of a calculation and you do not use parentheses correctly you may inadvertently enter a value very different from what you intended. Instead, you should enter this number by pressing "6" and then pressing the "EXP" or "EE" button, followed by "24". That's because the "EXP" or "EE" button on scientific calculators means "times 10 to the power of". So entering 6 EXP 24 or 6 EE 24 stores the number 6 times 10 to the power of 24 in the calculator.

Our warning from Section 1.4.1 is worth repeating here: If you see a number in scientific notation that has no coefficient at all, such as 10^6, it is important to remember that a coefficient of 1.0 is implicit. That is, $10^6 = 1 \times 10^6$. To enter this in your calculator, type 1 EXP 6. Many students go astray and type 10 EXP 6, which is incorrect because it represents 10×10^6, or 10^7.

If you're thinking "I've always entered numbers by pressing the × button and then pressing 10 and ^, and I've been getting the right answers," you should understand that this approach may give a correct or an incorrect answer depending on exactly which buttons you press before and after entering a number this way. Worst of all, you won't know in any given case if you've managed to get the correct number into the calculator this way. Since it takes just a few minutes to learn to use the EXP or EE buttons to enter numbers in scientific notation, and doing so will save you time and mistakes down the road, why take the risk that the other approach may fail you at a critical time?

One question that many students have about astronomy calculations is how many digits should be used in entering numbers into, and reporting answers from, calculators. This gets into the question of significant figures, for which you can find links to helpful resources on this book's website. But the short answer is that most astronomy texts present values accurate to two decimal places (at most), so reporting your answers with one or two decimal places will suffice for most problems.[7] However, this does not mean that you should round your intermediate results to two digits – the best practice is to keep lots of digits (6 or 8) *during* your calculation, and then round only when reporting your final answer. This will minimize the build-up of rounding errors during multi-step calculations.

1.5 Chapter problems

1.1 Express 1.5×10^{17} seconds (the remaining lifetime of our Sun) in years.

1.2 The acceleration of gravity at the Earth's surface is 9.8 m/s^2. Convert this number to ft/s^2 and miles/hr^2.

1.3 It takes you 40 minutes to walk 2 miles to work. What is your average speed in miles per hour? Explain how you know your answer makes sense.

1.4 We communicate with spacecraft using radio wavelengths of light, which travel at speed $c = 3 \times 10^8$ m/s. The distance between Earth and Mars varies between 56 million and 400 million kilometers.

(a) Without using a calculator, estimate how long the signal takes to reach Mars when it is 300 million kilometers away using the absolute method.

(b) Now, using the absolute method with a calculator, calculate how much time the signal takes to reach Mars at closest approach.

[7] There are some notable exceptions such as Doppler-shift problems, which you can read about in Section 3.3

(c) Using your answer to the previous part and the ratio method, calculate the travel time for the signal to reach Mars at its farthest distance.

1.5 A fast runner can complete a 1-mile (1,600-m) race in 4 minutes. At this speed, how long would it take to run the distance to the Moon $(3.8 \times 10^5$ km)?

1.6 The radii of Earth, Jupiter, and the Sun are $R_E = 6,371$ km, $R_J = 69,911$ km, and $R_S = 696,000$ km, respectively.

(a) How many Earths could fit in Jupiter?

(b) How many Earths could fit in the Sun?

1.7 The "light-gathering power" of a telescope is directly proportional to the area of the telescope's lens or mirror, and more light-gathering power lets you see fainter objects. Compare the light-gathering power of a telescope with a lens with diameter of 4 inches with the light-gathering power of a human eye with pupil diameter of 6 mm.

1.8 A galaxy is a gigantic collection of stars. If there are a hundred billion stars per galaxy, a hundred billion galaxies in the observable Universe, and two planets per star, how many planets are there in the observable Universe?

1.9 The force of Earth's gravity on an object is inversely proportional to the square of the object's distance from the center of the Earth. Compare the force of Earth's gravity on the Voyager spacecraft when it was just leaving Earth's atmosphere (at a distance of about 6,450 km from the Earth's center) to the force of Earth's gravity on Voyager at its current distance of about 2×10^{10} km.

1.10 The Hubble Ultra-Deep Field (HUDF) is a famous photograph of a tiny portion of the sky that captured images of about 10,000 galaxies over an exposure time of one million seconds. At this rate, how many years would it take to photograph all of the estimated 100 billion galaxies in our observable Universe?

2

Gravity

Even before taking an astronomy class, most people have a sense of how gravity works. No mathematics is needed to understand the idea that every mass attracts every other mass and that gravity is the force that causes apples to fall from trees. But what if you want to know how much you'd weigh on Saturn's moon Titan, or why the Moon doesn't come crashing down onto the Earth, or how it can possibly be true that you're tugging on the Earth exactly as hard as the Earth is tugging on you? The best way to answer questions like that is to gain a practical understanding of Newton's Law of Gravity and related principles.

This chapter is designed to help you achieve that understanding. It begins with an overview of Newton's Law of Gravity, in which you'll find a detailed explanation of the meaning of each term. You'll also find plenty of examples showing how to use this law – with or without a calculator. Later sections of this chapter deal with Newton's Laws of Motion as well as Kepler's Laws. And like every chapter in this book, this one is modular. So, if you're solid on gravity but would like a review of Newton's Third Law, you can skip to that section and dive right in.

2.1 Newton's Law of Gravity

The equation for Newton's Law of Gravity may look a bit daunting at first but, like most equations, it becomes far less imposing when you take it apart and examine each term. To help with that process, we'll write "expanded" versions of some of the important equations in this book, of which you can see an example in Figure 2.1. As you can see, in an expanded equation, the meaning and units of each term are readily available in a text block with an arrow pointing to the relevant term. After the figure, you'll find additional explanations of the

Gravity

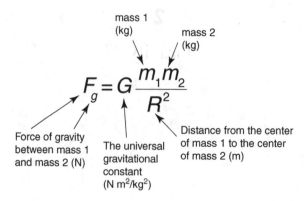

Figure 2.1 Newton's Law of Gravity.

terms as well as examples of how to apply the equation using both the absolute method and the ratio method.

2.1.1 Description of terms in the gravity equation

Whenever you encounter an equation such as that shown in Figure 2.1, it's a good idea to make sure you understand not only the meaning (and units) of each term, but also what the placement and powers of those terms are telling you.

The force of gravity, F_g, appears on the left side of this equation in units of newtons (N). The force occurs between two objects such as those shown in Figure 2.2; each object produces the gravitational force F_g on the other. Here are detailed descriptions of each of the terms on the right side of the gravity equation:

G The first term is G, the universal gravitational constant.[1] To the best of scientists' knowledge, this constant has the same value throughout the known Universe, and that value in SI units is $6.67 \times 10^{-11} \text{N m}^2/\text{kg}^2$.

m_1, m_2 The variables in the numerator of the fraction, m_1 and m_2, represent the amount of mass (in units of kilograms) in each of the two objects for which the force of gravity is being calculated. Most astronomy texts use lowercase "m," as we have here, as a variable (a placeholder for an unspecified quantity) to represent mass. Be wary that this invites confusion with the

---

[1] This constant is always written with an uppercase G – do not confuse this with lowercase g, which is usually used to denote the acceleration produced by gravity at a specific location. G and g have different units and different meanings.

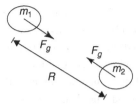

Figure 2.2 Two masses tugging on each other.

abbreviation for the distance unit meters, which is also abbreviated with a
lowercase "m." Take care never to confuse the two. Some texts, like this one,
italicize variables but not units – so "m" would represent a variable for mass,
and "m" would be an abbreviation for the unit meters – but convention varies
between texts, so you may have to scrutinize your text to ascertain whether
or not it uses this convention. Moreover, if there's a subscript after the m,
that's a strong hint that m represents a variable for the quantity mass, not the
units meters – but again, convention varies between texts. You may have to
judge from context which meaning "m" has in different circumstances.

   You should note that the mass of an object is a measure of the total amount
of material that makes up the object and is *not* the same as the weight of the
object. As you will see in the first example below, the weight of an object
is simply the force of gravity (usually expressed in pounds in everyday life,
rather than newtons), and that force depends on exactly where the object
is located. So, your weight on the surface of the Earth is greater than your
weight on the surface of the Moon because the Moon produces a smaller
force of gravity at its surface. But your mass is the same no matter where
you are.

R It's quite common for students to assume that the R in the denominator of
   the gravity equation means the "radius" of an object, but in fact it represents
   the *distance* (in units of meters) between the center[2] of mass 1 and the center
   of mass 2. There are certainly some cases in which the distance R turns out
   to be approximately equal to the radius of a sphere (such as a planet), but
   you should not fall into the habit of thinking of R as always being a radius.

Once you're comfortable with the meaning and units of each term, it's time
to step back and consider what the placement and power of those terms is
telling you about the force of gravity. The fact that both masses appear in the

---

[2] The word "center" in this context means center of gravity, but for simplicity most astronomy
   texts assume spherically symmetric objects, for which the geometric center and the center of
   gravity are the same thing.

numerator on the right side of the equation tells you that the force of gravity is directly proportional to each of the masses. So, if you double one of the masses while keeping everything else the same, the force of gravity between the masses will also double. Notice also that it doesn't matter which mass you call m_1 and which you call m_2; since multiplication is commutative, they are interchangeable. And since you calculate only one force F_g between the two masses, the force of gravity of m_1 on m_2 is exactly the same as the force of gravity of m_2 on m_1. This is an example of Newton's Third Law (which you can read about later in this chapter) and it means that right now you're pulling on the Earth exactly as hard as the Earth is pulling on you.

Now consider the placement and power of the R term in the gravity equation. Since the distance between the objects appears in the denominator, that means force and distance squared are inversely proportional, so greater distance will result in smaller gravitational force (exactly as you would expect, since common sense tells you that nearby objects exert a greater gravitational pull than far-away objects). Since R is squared, that means that the force of gravity drops off rapidly with distance. So doubling the distance while keeping everything else the same does not cause the force to decrease to one-half its original value (as it would if distance appeared to the first power in the denominator). Instead, doubling the distance reduces the force to one-quarter its original value (since $\frac{1}{2^2} = \frac{1}{4}$). This is called the "inverse-square" law relating force and distance – inverse because of the inverse proportionality, and square because the R term is raised to the second power.

With an understanding of the meaning of the gravity equation, you're ready to use this equation to solve astronomy problems. As described in Section 1.2, there are two ways to use an equation like this to solve problems. The absolute method can be used to find the value of the force of gravity (in newtons) by plugging numerical values into the gravity equation. The ratio method is useful if you wish to compare the force of gravity between objects under two different sets of circumstances. You can see an example of the use of that approach a bit later in this section.

2.1.2 Calculating the force of gravity

Using the absolute method, you enter the values of all variables (in this case, the masses of m_1 and m_2 in kilograms and the distance in meters) and constants (here, only G) in appropriate units. If you're given the values of any variables in different units, you'll need to convert to the required units. Then perform the necessary mathematical operations to arrive at an "absolute" answer – that is, an answer that represents a value with appropriate units rather than a relative answer. This is the approach to use if you're trying to find the force of gravity between two objects of known mass at a known distance. Here's an example:

Example: Calculate the force of gravity between the Sun and the planet Uranus.

A good way to begin any problem is to write down exactly what you're given, what you're trying to find, and what relationship connects what you're given to what you're trying to find.

In this case, you're given the names of two objects (the Sun and Uranus), and you're asked to find the force of gravity between them. You know that Newton's Law of Gravity can be used to find the force of gravity between any two objects, as long as you know the mass of each object and the distance between them. And although the problem statement doesn't give you the mass of either the Sun or Uranus or the distance between them, you can find that information in most comprehensive astronomy texts or on-line.

Using those resources, you should be able to find that the mass of the Sun is about 2×10^{30} kg, the mass of the planet Uranus is about 8.7×10^{25} kg, and Uranus's distance from the Sun varies from about 2.74×10^{9} to 3.01×10^{9} km. Since the problem doesn't specify the point in Uranus's orbit at which you should find the force of gravity, you're free to use either of those values or something in between. If you take the middle of that range (2.87×10^{9} km) as the distance, you have all the quantities needed to find the force of gravity. But before you can start plugging values into Newton's Law of Gravity, it's essential that you remember to convert the distance into the required units of meters:

$$R = 2.87 \times 10^{9} \text{ km} \times \left(\frac{1,000 \text{ m}}{1 \text{ km}} \right) = 2.87 \times 10^{12} \text{ m}.$$

Now you can plug in the values for the masses and distance, like this:

$$F_g = G \frac{m_1 m_2}{R^2} \qquad (2.1)$$

$$= \left(6.67 \times 10^{-11} \frac{\text{N m}^2}{\text{kg}^2} \right) \left[\frac{(2 \times 10^{30} \text{ kg})(8.7 \times 10^{25} \text{ kg})}{(2.87 \times 10^{12} \text{ m})^2} \right]$$

$$= \left(6.67 \times 10^{-11} \frac{\text{N m}^2}{\text{kg}^2} \right) \left(2.11 \times 10^{31} \frac{\text{kg}^2}{\text{m}^2} \right)$$

$$= 1.4 \times 10^{21} \text{ N}.$$

Exercise 2.1. Calculate the force of gravity between two people, each with mass of 80 kg, if the distance between them is 2 meters.

2.1.3 Surface gravity

One very common (and practical) type of gravity problem in astronomy is to determine the force of gravity between a celestial body (such as a moon, planet, or star) and an object on the surface[3] of that body. According to Newton's Law of Gravity, that force depends on the mass of the celestial body, the mass of the object on the surface, and the distance from the center of the body to the center of the object.

To understand the relevant distance, consider the case of a person standing on the surface of a planet, as shown in Figure 2.3. As suggested by this figure, the distance from the center of the planet to the center of the person is very well approximated by the radius of the planet (since the planet's radius is typically thousands of kilometers and the person's height is 2 meters or less). So in this case the "R" term in the denominator of Newton's Law of Gravity turns out to be the radius of the body to a very good approximation.

This means that if you (with mass m_{you}) are standing on the surface of a planet (with mass m_{planet}), the force of gravity (F_g) between you and the planet is

$$F_g = G \frac{m_{you} m_{planet}}{R^2_{planet}}.$$

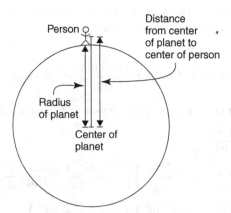

Figure 2.3 Radius of planet and center-to-center distance.

---

[3] Some objects, such as stars and gas-giant planets, lack a well-defined surface. For stars, the "surface" is often defined as the photosphere of the star (that's the layer from which the star's light radiates), although no solid surface exists at that location. Likewise, for gas-giant planets, the "surface" level is sometimes associated with a layer in the planet's atmosphere.

Here's an example:

Example: Find the force of gravity of the Earth on a person with a mass of 100 kg standing on the Earth's surface.

In this case, the only number you're given is the mass of the person (100 kg), which you can call m_1. But you're also told that the person is standing on the surface of the Earth, which means that you can look up the mass of the second object (the Earth), which you can call m_2, and the distance between the person and the center of the Earth (R), which is essentially the radius of the Earth. In any comprehensive astronomy text, you should be able to find[4] the mass of the Earth as about 6×10^{24} kg and the radius of the Earth as approximately 6,378 km. So, you have the values of two masses (the person and the Earth) as well as the distance between their centers.

Clearly, the relationship between two masses, the distance between their centers, and the force of gravity between them is provided by Newton's Law of Gravity. So you can solve this problem by using the gravity equation, but first you're going to have to make sure that the variables have the units required by that equation. You have the mass of the person and the mass of the Earth in kilograms, as required, but the distance between their centers is given in kilometers rather than meters. That's an easy conversion (see Section 1.1 if you need help with unit conversions); multiplying 6,378 km by the conversion factor of 1,000 meters per kilometer gives you the value of R: 6,378,000 m, or 6.378×10^6 m (see Section 1.4 if you need help with scientific notation).

Inserting these values along with the constant G into the gravity equation gives

$$F_g = G\frac{m_1 m_2}{R^2} \tag{2.1}$$

$$F_g = \left(6.67 \times 10^{-11}\,\frac{\text{N m}^2}{\text{kg}^2}\right)\left[\frac{(100\ \text{kg})(6 \times 10^{24}\ \text{kg})}{(6.378 \times 10^6\ \text{m})^2}\right]$$

$$= \left(6.67 \times 10^{-11}\,\frac{\text{N}\,\cancel{\text{m}^2}}{\cancel{\text{kg}^2}}\right)\left(1.475 \times 10^{13}\,\frac{\cancel{\text{kg}^2}}{\cancel{\text{m}^2}}\right)$$

$$= 983.8\ \text{N},$$

which is the gravitational force between the Earth and a 100-kg person standing on the Earth's surface. To put this into more-familiar terms, you can convert the units of this result from newtons to pounds by using the conversion factor

---

[4] If you don't know where to find it, use the index.

of 1 lb ↔ 4.45 N. This means that the force of Earth's gravity in pounds on a 100-kg person is

$$F_g = (983.8\,\text{N})\frac{1\,\text{lb}}{4.45\,\text{N}} = 221\,\text{lb}.$$

As always, you should ask yourself whether this answer makes sense. And it does! Since 1 kg "equals" 2.2 lb, it makes sense that 100 kg "equals" about 220 lb. But can 1 kg (which is an amount of mass) really "equal" 2.2 lb (which is a force)? Clearly not, which is why we put quotation marks around the word "equals" in the previous sentences. The reason it's commonly said that 1 kg equals 2.2 lb is that *on the surface of the Earth* the force of Earth's gravity on a mass of 1 kg is about 2.2 lb. But take that 1-kg mass just about anywhere else in the Universe, and the force of gravity on it will not be 2.2 lb. To verify that, you can use the gravity equation to find the force of gravity produced by Earth's Moon on a mass on the surface of the Moon.

Example: Find the force of the Moon's gravity on a 1-kg object on the surface of the Moon.

The Moon's mass is about 7.35×10^{22} kg, and its radius is about 1,737 km, so the gravity equation gives

$$F_g = G\frac{m_1 m_2}{R^2} \tag{2.1}$$

$$F_g = \left(6.67 \times 10^{-11}\frac{\text{N m}^2}{\text{kg}^2}\right)\frac{(1\,\text{kg})(7.35 \times 10^{22}\,\text{kg})}{(1.737 \times 10^6\,\text{m})^2}$$

$$= 1.62\,\text{N},$$

which is about 0.37 lb. This result means that 1 kg does not "equal" 2.2 lb on the surface of the Moon. Instead, on the surface of the Moon, a mass of 1 kg weighs only about $\frac{1}{3}$ lb.

Since 0.37 is very close to one-sixth of 2.2, it's frequently said that you would weigh only about one-sixth as much on the Moon as on Earth, or "the Moon's gravity is one-sixth of Earth's gravity".

It is precisely this kind of comparison that is easily done using the ratio method described in Chapter 1. Whenever you're asked to compare quantities (often through questions such as "How many times bigger . . . ?" or "How much stronger . . . ?"), you should consider using the ratio method. But even when you're trying to find an absolute quantity such as the force of gravity on a certain planet, as long as you have a reference value (such as the force of gravity on Earth), you may still be able to save a lot of time and effort – and minimize what you have to plug into your calculator, thus reducing chances for errors – by using ratios where possible.

You can find a detailed description of and motivation for the ratio method in Section 1.2, but here's a quick summary of how to apply it to an example gravity problem.

Example: Compare the force of Earth's gravity on an object on Earth's surface with the force of the Moon's gravity on the same object on the surface of the Moon.

If you use the ratio method, it's not necessary to go through the process of finding the value of F_g on the Earth, then finding the value of F_g on the Moon, and then dividing one by the other (if you're wondering why it's often better to compare quantities in astronomy by dividing rather than by subtracting, that's also discussed in Section 1.2).

You can certainly get the correct answer that way, but you're taking a lot of steps that can be avoided by realizing that quantities of the mass of the object (call it m_1) and G will be exactly the same in your calculations for both locations (Earth and Moon).

By not plugging in values until the very end, you'll see that m_1 and G will cancel out.

This may be a little more clear if you consider what's going to happen when you take that crucial "compare by dividing" step. If you call the force of gravity on Earth $F_{g,Earth}$ and the force of gravity on the Moon $F_{g,Moon}$, dividing will look like this:

$$\frac{F_{g,Moon}}{F_{g,Earth}} = \frac{G\frac{m_1 m_{Moon}}{R_{Moon}^2}}{G\frac{m_1 m_{Earth}}{R_{Earth}^2}} = \frac{\cancel{G}\frac{\cancel{m_1} m_{Moon}}{R_{Moon}^2}}{\cancel{G}\frac{\cancel{m_1} m_{Earth}}{R_{Earth}^2}} = \frac{\frac{m_{Moon}}{R_{Moon}^2}}{\frac{m_{Earth}}{R_{Earth}^2}} = \frac{m_{Moon}}{R_{Moon}^2} \times \frac{R_{Earth}^2}{m_{Earth}}$$

$$= \left(\frac{m_{Moon}}{m_{Earth}}\right)\left(\frac{R_{Earth}^2}{R_{Moon}^2}\right) = \left(\frac{m_{Moon}}{m_{Earth}}\right)\left(\frac{R_{Earth}}{R_{Moon}}\right)^2 . \tag{2.2}$$

Now plugging in values for the masses of Earth and Moon, and the distances between their centers and surfaces (which are equal to their radii), gives

$$\frac{F_{g,Moon}}{F_{g,Earth}} = \left(\frac{m_{Moon}}{m_{Earth}}\right)\left(\frac{R_{Earth}}{R_{Moon}}\right)^2 = \left(\frac{7.35 \times 10^{22} \text{ kg}}{6 \times 10^{24} \text{ kg}}\right)\left(\frac{6,378 \text{ km}}{1,737 \text{ km}}\right)^2$$

$$= 0.165 \approx \frac{1}{6} .$$

This is the same result you obtained earlier using the absolute method to calculate the force of the Moon's gravity on a 1-kg mass, convert that force from newtons to pounds and finally to compare it to the weight of a 1-kg mass on Earth (2.2 lb). But in the ratio approach you didn't have to enter a value for m_1 because it cancelled. In other words, the result is *independent of mass m_1* – the

Moon's gravity is one-sixth of Earth's gravity for an object of any mass. Furthermore, notice that this time you didn't have to do a unit conversion to get distance into units of meters. Since you had both distances in kilometers, when you divided them the units cancelled. So with the ratio method the quantities can be in any units, as long as the units are the same in the two scenarios you are comparing.

The simplicity of this result dramatically illustrates the power of the ratio method. To compare the Moon's gravity to the Earth's, just look at Eq. 2.2. This equation says that to compare $F_{g,Moon}$ to $F_{g,Earth}$, you can simply divide the mass of the Moon by the mass of the Earth and multiply that result by the square of the ratio of the Earth's radius to the Moon's radius. In terms of calculator operations, the four number entries, two divides, one squaring, and one multiply (8 total entries) of the ratio method replace the ten number entries, five divides, two squarings, and four multiplies (21 total entries) of the absolute method, and that's not counting unit conversions.

The power of the ratio method is even more compelling in cases in which you're already given values in ratio form, as you can see in the following example.

Example: Compare Earth's surface gravity to that of Jupiter, whose radius is 11.2 times the radius of the Earth and whose mass is 318 times the mass of Earth.

In such a problem, there's no need to calculate the force of gravity on a certain mass (m_1) on Jupiter and on Earth and then divide one result by the other. Just as in the previous example, m_1 and G will cancel so you can simply start with the Jupiter/Earth version of Eq. 2.2:

$$\frac{F_{g,Jupiter}}{F_{g,Earth}} = \left(\frac{m_{Jupiter}}{m_{Earth}}\right)\left(\frac{R_{Earth}}{R_{Jupiter}}\right)^2 = (318)\left(\frac{1}{11.2}\right)^2 = 2.5,$$

or $F_{g,Jupiter} = 2.5 \, F_{g,Earth}$. Translating this mathematical result into a sentence, this means that the force of gravity on the "surface" of Jupiter is 2.5 times stronger than the force of gravity on the surface of Earth (for help interpreting ratio answers, see Section 1.2.4).

If it seems strange to you that the force of gravity on the surface of a planet that's over 300 times more massive than Earth is only 2.5 times greater than the force of gravity on the surface of the Earth, remember the $\frac{1}{R^2}$-term in the denominator of Newton's Law of Gravity. Because Jupiter is much larger than Earth, the distance from the "surface" to the center of Jupiter is much greater than the distance from the surface to the center of Earth, and this greater distance partially compensates for the greater mass of Jupiter.

You can get some practice doing gravity problems by working through the following exercise, and you'll find several gravity problems in the problem set at the end of this chapter.

Exercise 2.2. Calculate the force of gravity (that is, the weight) in newtons of a 50-kg person on the surfaces of the following three planets.

(a) **Earth (R_{Earth} = 6378 km and m_{Earth} = 6 × 10^{24} kg).**

(b) **Mars (R_{Mars} = $0.53 R_{Earth}$, and m_{Mars} = $0.11 m_{Earth}$).**

(c) **Saturn (R_{Saturn} = $9.5 R_{Earth}$, and m_{Saturn} = $95.2 m_{Earth}$).**

(d) **Now convert those weights from newtons to pounds, using the conversion factor 1 lb ↔ 4.45 N.**

2.2 Newton's Laws of Motion

If you've ever taken a physics class, you've almost certainly seen some form of Newton's Laws of Motion. However, not all physics classes impart an intuitive feel for what these laws mean or their implications on the behavior of celestial objects. Some astronomy classes spend little or no time explicitly discussing Newton's Laws, while others mention these laws as the basis of the science of Mechanics and the "why" behind Kepler's Laws (which you can read about in Section 2.3). So, although you may not run into many problems explicitly involving Newton's Laws of Motion, you'll have a much better understanding of the way the Universe works if you comprehend these laws (and some professors do expect you to understand them). That's why this section will briefly review the relevant concepts and show you how to solve problems using Newton's Laws of Motion.

Newton's Laws of Motion are usually presented using words such as these:

First Law   An object at rest will remain at rest unless an unbalanced force[5] acts upon it, and an object in motion will continue moving at the same speed and in the same direction unless an unbalanced force acts upon it.

Second Law   If an unbalanced force (F) acts upon an object of mass (m), the object will experience an acceleration (a) given by

$$a = \frac{F}{m}. \tag{2.3}$$

---

[5] An unbalanced force is the force left over when you add up all of the forces acting on an object, some of which may partially or fully cancel one another. The unbalanced force is also called the net force or the total force.

This equation is often written as $F = ma$, and it applies only to situations in which the mass does not change during the acceleration.

Third Law If one object produces a force on a second object, the second object produces an equally strong force back onto the first object in the opposite direction. These are sometimes called "equal and opposite forces," and it's very important that you remember that these two forces act on *two different objects*.

The First Law is sometimes called "the law of inertia" because the word "inertia" means resistance to acceleration. This law is simply a statement that an object will not accelerate (that is, its velocity will not change) unless an unbalanced force is applied to it. And since velocity is a vector that combines speed and direction, a change of velocity (that is, an acceleration) can mean speeding up, slowing down, or turning. So, if no force is acting on an object, or if the forces all cancel, the object will not speed up, it will not slow down, and it will not turn.

Many students find Newton's First Law counter-intuitive, believing that a force is needed to cause an object to continue moving and that removing all forces will cause an object to slow down and eventually come to a halt (even Aristotle believed that the "natural state" of solid objects is to be at rest). It's not hard to understand why many people have that mistaken impression, because we live under the constant influence of Earth's gravitational force (which tends to pull objects to the ground) and a host of frictional forces such as air resistance and surface friction (which tend to cause moving objects to slow down).

But Newton's First Law tells you that if you throw a 90-mile-per-hour (mph) fastball out of the window of your spaceship in the vacuum of outer space (far away from gravitational and frictional forces), that ball will continue traveling at 90 mph in the same direction forever if no unbalanced force acts upon it. Understanding this law can help you answer a question often posed by astronomy students: what would happen to an orbiting object (such as the Earth) if the object being orbited (such as the Sun) were suddenly to disappear? Assuming that the disappearing object took all of its mass with it (something that never happens in nature), the unbalanced gravitational force on the orbiting object would suddenly become zero, and from that instant onward it would not speed up, slow down, or turn. Suddenly "untethered," the object would sail off at the same speed and in the same direction it was moving at the instant the gravity disappeared – that direction is not radially outward from the center of its orbit, but tangential to the orbit.

Newton's Third Law (sometimes called the action/reaction law) doesn't require much mathematics to apply. If you apply a force to the top of a desk by slamming your hand onto it, the desk produces a force on your hand that is exactly as strong but in the opposite direction – that's why your hand hurts if you do this. So, if you use Newton's Law of Gravity to calculate the force that mass m_1 produces on mass m_2, that's also the force that m_2 produces on m_1. That means that right now you're pulling on the Earth just as hard as the Earth is pulling on you. This is relevant, for example, in the detection of planets around other stars. The mutual gravity between a planet and its star makes the planet move around in its orbit, but it also makes the star move in a small orbit in response, as described in Section 3.4.

Newton's Third Law causes some students to wonder why you accelerate toward the ground (rather than the ground rising up to meet you) when you fall out of a tree. To understand that, you need to consider the role of mass in Newton's Second Law.

That role can be understood by looking at the position of the mass term (m) in Eq. 2.3. Since mass is in the denominator, the acceleration (a) produced by a certain force (F) will be smaller for large masses than for small ones. In other words, acceleration is *inversely* proportional to mass (but directly proportional to force, since force is in the numerator). That means that, for a given mass, doubling the force will produce twice the acceleration, but for a given force, doubling the mass will produce half the acceleration. So although you're pulling on the Earth with just as strong a force as the Earth is pulling on you, your acceleration is much greater than Earth's because your mass is much less than Earth's.

This is an important demonstration of the best way to think about equations – not as something useless until numbers are plugged in, but as powerful and precise statements of the relationship between quantities. So if you want to know the relationship between acceleration, force, and mass, Newton's Second Law provides the answer.

One very useful application of Newton's Second Law is to determine the acceleration of an object (such as a person) produced by another object (such as a planet) through gravity. You can see how this works in the following example:

Example: Find the gravitational acceleration of an object of mass m_1 produced by another object (of mass m_2) at a distance R.

Using Newton's Second Law, you know the acceleration of m_1 will be given by $a = F/m_1$, where F is the total force acting on mass m_1. But you also know that the force of gravity acting between m_1 and m_2 is given by

$$F_g = G\frac{m_1 m_2}{R^2}, \tag{2.1}$$

where G is the universal gravitational constant and R is the distance between the centers of m_1 and m_2.

Using F_g as the force in Newton's Second Law gives

$$a = \frac{F_g}{m_1} = \frac{G\frac{m_1 m_2}{R^2}}{m_1} = \frac{G\frac{\cancel{m_1} m_2}{R^2}}{\cancel{m_1}} = G\frac{m_2}{R^2}. \tag{2.4}$$

So the acceleration of m_1 due to the force of gravity depends only on the mass of the other object (m_2) and its distance – not on the amount of mass of the object itself (m_1). That's why, in the absence of other forces such as air resistance, all objects fall to Earth with the same acceleration regardless of their mass.

How can this possibly be true? Shouldn't an object with bigger mass fall faster, since the gravitational force between that object and Earth is greater?

Well, it's true that objects with greater mass experience a greater gravitational force from the Earth, but remember Newton's Second Law: objects with greater mass *resist acceleration* more than objects with less mass (that is, more-massive objects have greater inertia). So, although the gravitational force is directly proportional to mass, acceleration is inversely proportional to mass, and combining these dependencies means that the acceleration of an object due to gravity does not depend on the object's mass.

To find the acceleration of an object near the surface of the Earth, just plug the values for the Earth's mass and radius into Eq. 2.4:

$$a = G\frac{m_{Earth}}{R_{Earth}^2} = \left(6.67 \times 10^{-11} \frac{N\cancel{m^2}}{\cancel{kg^2}}\right)\left[\frac{6 \times 10^{24}\cancel{kg}}{(6.38 \times 10^6 \cancel{m})^2}\right]$$

$$= 9.8\frac{m}{s^2}.$$

If you don't see how the units work out in this equation, remember that newtons are equivalent to kg m/s^2, as discussed in Section 1.1.6.

So, near the surface of the Earth, all objects experience the same acceleration due to gravity. In some texts, this gravitational acceleration is called g, which you should be careful not to confuse with G, the universal gravitational constant.

Here are some exercises to check your understanding of Newton's Laws of Motion; you'll find additional problems at the end of this chapter.

Exercise 2.3. How much force is needed to cause an automobile with mass of 1,200 kg to accelerate at a rate of 0.25 m/s^2?

Exercise 2.4. How many pounds of force does a 100-kg piano exert on the floor upon which it is resting? Compare this to the force of the floor upon the piano.

Exercise 2.5. Find the gravitational acceleration at the surface of Mars, which has a radius of 3,390 km and mass of 6.4×10^{23} kg.

2.3 Kepler's Laws

As with Newton's Laws of Motion, the basic concepts of Kepler's Laws of Planetary Orbits can be understood with a minimum of mathematics. But if you want to apply Kepler's Laws to problems in orbital dynamics, you should make sure you understand the mathematical underpinnings of these laws. The goal of this section is to help you achieve that understanding.

Kepler's Laws of Planetary Orbits are generally expressed in statements such as these:

First Law   The shape of a planet's orbit is an ellipse with the Sun at one focus, so the distance between a planet and the Sun is not, in general, constant.

Second Law   An imaginary line between a planet and the Sun sweeps out equal areas in equal times, so a planet moves faster when it's in the portion of its orbit closer to the Sun.

Third Law   The square of a planet's orbital period is proportional to the cube of the semi-major axis of the planet's orbit, so planets far from the Sun take longer to complete one orbit than planets close to the Sun.

Most introductory astronomy problems involving Kepler's Laws are based on Kepler's Third Law, but you may also encounter problems involving Kepler's First Law. If you do, it's likely that those problems involve the aphelion, perihelion, semi-major axis, and eccentricity of an orbit. To work those problems, you should begin by making sure you understand the meaning of those terms and their relationship to one another. The next two sections can help with that.

2.3.1 Ellipse parameters

Several of the basic parameters of an ellipse are shown in Figure 2.4. Remember that an ellipse is defined by two points called the "foci" of the ellipse (each one is called a "focus"). For all points on the ellipse, the combined distances from both foci have the same value. This is why you can draw an ellipse by inserting a thumbtack at each focus, attaching a string (loosely) between them,

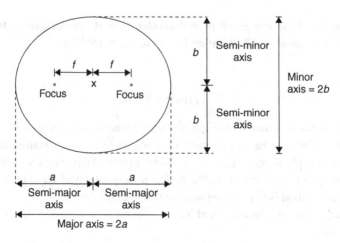

Figure 2.4 Parameters of an ellipse relevant to Kepler's First Law.

and then stretching the string tight with the tip of your writing instrument and moving all the way around both foci.

As long as you don't put your foci on top of one another, the resulting figure will have a long (major) axis and a shorter (minor) axis, with the length of the string equal to the major axis. As you can see in Figure 2.4, half the major axis is called the semi-major axis and is usually denoted "a" while half the minor axis is called the semi-minor axis and usually denoted "b." The distance from the center of the ellipse (marked with an "x") to each focus is sometimes called "f", so the distance between the two foci is equal to $2f$.

Unlike circles, which can have different sizes but all of which have the same shape, ellipses can have different shapes as well as different sizes. To see that, take a look at the four ellipses in Figure 2.5. Each of these ellipses has the same length semi-major axis, but their shapes are clearly not the same. If you're thinking that the top-left shape in Figure 2.5 isn't an ellipse at all, remember that a circle is just a special case of an ellipse, much like a square is a special case of a rectangle.

What's different about these ellipses is their flatness, which is called the "eccentricity" of the ellipse. The eccentricity (e) is defined as

$$e = \sqrt{1 - \frac{b^2}{a^2}} = \frac{f}{a},\qquad(2.5)$$

where a is the semi-major axis, b is the semi-minor axis, and f is the distance from the center of the ellipse to either focus. In this equation, a, b, and f must all have the same units.

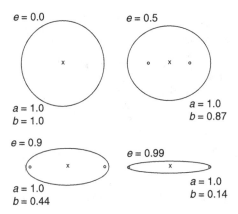

Figure 2.5 Ellipses of different eccentricities.

As you can see from this equation, a circle (for which $a = b$) has eccentricity (e) of zero. For a highly flattened ellipse (a much greater than b), the eccentricity approaches one. The foci, centers, and values of the eccentricity, semi-major axis, and semi-minor axis are shown for each ellipse in Figure 2.5, so you can verify the values of the eccentricity using a ruler and Eq. 2.5 if you're interested.

2.3.2 Elliptical orbit parameters

In astronomy, the characteristics of ellipses can be used to analyze the paths that planets follow around the Sun. When you work such problems, you may come across the terms shown in Figure 2.6. As you can see, the Sun is at one focus of the ellipse, nothing is at the center or the other focus, and a planet is traveling along the ellipse as it orbits the Sun. This is only an approximation, because the star and planet are actually orbiting their common *center of mass*, which lies at the focus (as described in Section 3.4). However, because a planet's mass is usually negligible compared to the mass of a star, in response to the equal force of gravity on both objects the star's acceleration (and hence its orbit) will be minuscule compared to the planet's huge acceleration (and larger orbit), so it is a very good approximation to treat the star as fixed with the planet orbiting it.

The point in the planet's orbit that lies closest to the Sun is called the perihelion, and the point farthest from the Sun is called the aphelion. From the figure, you can see that if the distance from the perihelion to the Sun is $dist_{peri}$ and the distance from the aphelion to the Sun is $dist_{ap}$, then it must be true that

$$dist_{peri} + dist_{ap} = 2a, \tag{2.6}$$

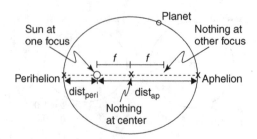

Figure 2.6 Parameters of an elliptical orbit.

where $2a$ is the major axis of the orbit. Solving this equation for a makes it clear that

$$a = \frac{dist_{peri} + dist_{ap}}{2},\qquad(2.7)$$

which means that the semi-major axis is equal to the average of the perihelion and aphelion distances.

Close inspection of Figure 2.6 also reveals the following relationships:

$$dist_{ap} = a + f,$$
$$dist_{peri} = a - f.\qquad(2.8)$$

You can find other useful relationships between orbital parameters by noting that since $e = f/a$ (Eq. 2.5), then $f = ae$, and Eqs. 2.8 become

$$dist_{ap} = a + f = a + ae = a(1 + e),$$
$$dist_{peri} = a - f = a - ae = a(1 - e),\qquad(2.9)$$

which are very helpful if you're trying to find the aphelion and perihelion distances when you know the semi-major axis (a) and eccentricity (e).

And here's a trick you may find useful when you're given two simultaneous[6] equations such as Eqs. 2.8 and you want to isolate one of the variables (such as a or e): try adding or subtracting the equations. Here's what happens when you add the two equations of Eq. 2.8:

$$dist_{ap} = a + f$$
$$+ \ \underline{dist_{peri} = a - f}$$
$$dist_{ap} + dist_{peri} = 2a + 0,$$

---

[6] Simultaneous equations are two or more independent equations that contain the same variables.

which verifies Eq. 2.6. Another useful relation can be found by subtracting the
two equations of Eqs. 2.8 and recalling from Eq. 2.5 that $f = ae$:

$$dist_{ap} \;= a + f$$
$$- \;\; dist_{peri} = a - f$$
$$dist_{ap} - dist_{peri} = 0 + 2f = 2ae.$$

Dividing both sides by $2a$ provides an expression for the eccentricity (e) in
terms of the aphelion and perihelion distances:

$$\frac{dist_{ap} - dist_{peri}}{2a} = \frac{2ae}{2a} = e, \tag{2.10}$$

and, since $2a = dist_{ap} + dist_{peri}$ (the long axis of the ellipse),

$$e = \frac{dist_{ap} - dist_{peri}}{2a} = \frac{dist_{ap} - dist_{peri}}{dist_{ap} + dist_{peri}}. \tag{2.11}$$

Since there are so many different ways to relate orbital parameters, we've
gathered the ones we find most useful into Table 2.1.

All of these relationships flow directly from Kepler's First Law stating that
planetary orbits are ellipses and Eqs. 2.5 and 2.6, which define the parameters

Table 2.1 *Orbit-parameter relationships*

To find	If you know	Use
$dist_{ap}$	a and e	$dist_{ap} = a(1 + e)$
$dist_{peri}$	a and e	$dist_{peri} = a(1 - e)$
a	$dist_{ap}$ and $dist_{peri}$	$2a = dist_{ap} + dist_{peri}$
e	$dist_{ap}$ and $dist_{peri}$	$e = \dfrac{dist_{ap} - dist_{peri}}{dist_{ap} + dist_{peri}}$
f	a and e	$f = ae$
f	a and b	$f = \sqrt{a^2 - b^2}$
f	$dist_{ap}$ and $dist_{peri}$	$2f = dist_{ap} - dist_{peri}$
e	a and b	$e = \sqrt{1 - \dfrac{b^2}{a^2}}$
e	a and f	$e = \dfrac{f}{a}$
b	e and a	$b = a\sqrt{1 - e^2}$
a	e and b	$a = \dfrac{b}{\sqrt{1 - e^2}}$

of those ellipses. Here are several exercises that will give you practice using these relationships to solve problems involving orbits.

Exercise 2.6. Mars' aphelion is about 1.67 AU and Mars' perihelion is about 1.38 AU. What is Mars' semi-major axis?

Exercise 2.7. What is the eccentricity of Mars' orbit?

Exercise 2.8. Earth's orbital eccentricity is approximately 0.017 and its semi-major axis is about 1.5×10^8 km. How far is the Sun from the center of Earth's orbit?

2.3.3 Using Kepler's Third Law

The vast majority of problems in introductory astronomy involving Kepler's Laws are based on Kepler's Third Law, which relates a planet's orbital period (P) to the semi-major axis (a) of its orbit. Kepler's Third Law is often written as

$$P^2 = a^3 \qquad (2.12)$$

but we prefer to write it as

$$[P \text{ (in yr)}]^2 = [a \text{ (in AU)}]^3 \qquad (2.13)$$

because this form explicitly reminds you that this equation works only if you express the period (P) in units of Earth years (yr) and the semi-major axis (a) in astronomical units (AU, where 1 AU = 149.6×10^6 km). You should also note that this form of Kepler's Third Law works only for objects orbiting the Sun.[7]

To see the importance of using the appropriate units when applying Kepler's Third Law, consider what would happen if you tried to insert Earth's orbital period as 365 days and the semi-major axis of Earth's orbit as 150 million kilometers into Eq. 2.12:

$$365^2 = (150 \times 10^6)^3, \qquad \text{(INCORRECT UNITS)}$$

which is clearly not true. But if you convert 365 days into 1 year and 150 million kilometers into 1 AU, Eq. 2.12 gives

$$1^2 = 1^3, \qquad \text{(CORRECT UNITS)}$$

which doesn't look nearly as impressive as the big numbers in the incorrect version, but this one has the distinct advantage of being true. So remember,

[7] Later in this section you can read about another form of Kepler's Third Law that works in other situations.

you can use $P^2 = a^3$ only when three conditions are met: (1) it's the Sun (or another object with the same mass as the Sun) that's being orbited, (2) the units of P are Earth years, and (3) the units of a are AU.

If you're wondering how the units of Kepler's Third Law work out, since P^2 has units of years squared and a^3 has units of AU cubed, the answer is that there's actually a hidden constant in this equation. With that constant shown explicitly, this equation would look like this:

$$P^2 = \left(1 \frac{\text{yr}^2}{\text{AU}^3}\right) a^3. \tag{2.14}$$

Notice that in addition to the strange combination of units, the constant has a numerical value of 1, which is why it's normally not written. Multiplying a^3 by this constant doesn't change your numerical result, but it does make the units balance between the left and right side of the equation.

Here's an example of how to use Kepler's Third Law to solve an orbital problem involving the planet Mars.

Example: Mars orbits the Sun once every 687 Earth days. What is the semi-major axis of Mars' orbit?

Since you're trying to find the semi-major axis a and you're given the period P, and since Mars is orbiting the Sun, this problem can be solved by applying Kepler's Third Law in the form

$$[P(\text{in yr})]^2 = [a(\text{in AU})]^3. \tag{2.13}$$

But before you can plug in your value for P and solve for a, you must convert 687 Earth days into the required units of Earth years. That's straightforward:

$$P = 687 \text{ days} = 687 \text{ days} \times \frac{1 \text{ yr}}{365 \text{ days}} = 1.88 \text{ yr.}$$

With P in the required units of Earth years, you can now use Kepler's Third Law to solve for the semi-major axis of Mars' orbit:

$$[P \text{ (in yr)}]^2 = [a \text{ (in AU)}]^3$$

$$a \text{ (in AU)} = \sqrt[3]{P \text{ (in yr)}^2} = ([P \text{ (in yr)}]^2)^{\frac{1}{3}}$$

$$= (1.88^2)^{\frac{1}{3}} = 3.53^{(\frac{1}{3})} = 1.52,$$

which means that Mars orbits the Sun at a distance of 1.52 AU, or about 50% farther than Earth does.[8] Notice that when you plug in numbers to do

---

[8] Be sure to include parentheses around the $\frac{1}{3}$ when entering this last step into your calculator lest it be interpreted as $(3.53^1)/3$, which will give you a wrong answer. If your calculator does not allow parentheses in exponents, then you can enter this as $3.53^{0.333}$.

calculations using Kepler's Third Law, it's not helpful to include the units. That's because this law as written in Eq. 2.13 explicitly shows the units for each quantity, so you know what those units must be. Furthermore, if you include the units, you'll also have to include the unwritten constant $\left(1\frac{yr^2}{AU^3}\right)$ for the units to work out properly. So this is one of very few cases in which it's not helpful to carry the units through every step of the calculation.

Since the version of Kepler's Third Law involving only P and a applies exclusively to objects orbiting the Sun, it's quite likely that you'll encounter a more universally applicable version of Kepler's Third Law. That version, devised by Isaac Newton, includes a mass term in the denominator:

$$P^2 = \frac{a^3}{M},\qquad(2.15)$$

or, as we like to write it to make the required units explicit:

$$[P(\text{in yr})]^2 = \frac{[a\ (\text{in AU})]^3}{M(\text{in solar masses})},\qquad(2.16)$$

where, as in Eqs. 2.12 and 2.13, P represents the orbital period and a represents the semi-major axis of the orbit. But this version of Kepler's Third Law has another term: M represents the mass of the object being orbited. This is an approximation because M is actually the sum of the masses of the orbiting object and the object being orbited. However, when one object is much more massive than the other such as a star and a planet, the planet's mass is negligible so it is a very good approximation to use the star's mass for M. If the masses of the two objects are similar to one another, as they may be in a binary star system, this approximation is invalid. Take care that M must be expressed in units of solar masses (masses of the Sun), where 1 solar mass $= 2 \times 10^{30}$ kg. Solar units are described in detail in Section 5.4.1.

Comparing Eq. 2.12 to Eq. 2.15, it may seem strange that these two equations could possibly both be correct. After all, Eq. 2.15 includes a term (M) that represents the gigantic masses of the Sun and a planet, and no such term appears in Eq. 2.12.

The key to understanding this apparent dichotomy is to consider the units of this term, shown explicitly as "solar masses" in Eq. 2.16. Since the Sun has a mass (by definition) of 1 solar mass, as long as you're considering a planet orbiting the Sun, you'll get the same answer from both versions of Kepler's Third Law (Eqs. 2.12 and 2.15), because dividing by 1 won't change your answer.

But it's very important that you remember that if you're working a problem involving a planet orbiting a star other than the Sun, or a moon orbiting a planet, or a satellite orbiting the Earth, then you must use the version of Kepler's Third Law containing mass in the denominator (Eqs. 2.15 or 2.16). Here's an example of such a problem.

Example: What is the orbital period of a telecommunications satellite in a circular orbit at a distance of 42,164 km from the center of the Earth?

In this problem, you're given the radius of the orbit, which is the same as the the semi-major axis a for a circular orbit. The body being orbited is not the Sun, so you cannot use Eq. 2.12; you must use Newton's modified version of Kepler's Third Law (Eq. 2.15 or 2.16). Begin by writing

$$[P \text{ (in yr)}]^2 = \frac{[a \text{ (in AU)}]^3}{M \text{ (in solar masses)}}. \tag{2.16}$$

But before you can begin plugging values into this equation, you must consider the units of the quantities you're given. For the semi-major axis (a) of the orbit, you're given the value of 42,164 km, but Eq. 2.16 specifies that the units of a must be AU, so a unit conversion is needed:

$$a = 42,164 \text{ km} = 42,164 \text{ km} \times \frac{1 \text{ AU}}{149.6 \times 10^6 \text{ km}} = 2.818 \times 10^{-4} \text{ AU}.$$

Now consider the mass term (M) in Eq. 2.15. It represents the combined mass of the satellite and the Earth, and you haven't been given the mass of the satellite. But even the heaviest satellites are billions of times less massive than the Earth, so the combined mass of the Earth and satellite is virtually identical to the mass of the Earth, which is about 6×10^{24} kg. Converting that to the required units of solar masses gives M as

$$M = 6 \times 10^{24} \text{ kg} = 6 \times 10^{24} \text{ kg} \times \frac{1 \text{ solar mass}}{2 \times 10^{30} \text{ kg}} = 3 \times 10^{-6} \text{ solar masses}.$$

With a and M in the required units, you're now ready to plug them in to Eq. 2.15 (again omitting units):

$$[P(\text{yr})]^2 = \frac{(2.818 \times 10^{-4})^3}{3 \times 10^{-6}} = 7.459 \times 10^{-6}$$

$$P(\text{yr}) = \sqrt{7.459 \times 10^{-6}} = 2.73 \times 10^{-3}.$$

A unit conversion reveals that 2.73×10^{-3} years is equivalent to 23 hours and 56 minutes, which is the same as the time it takes the Earth to rotate once on its axis. So satellites placed in a circular orbit above the equator at a distance of 42,164 km from the center of the Earth are "geosynchronous," which means that they orbit at the same rate that the Earth spins and remain over the same location on Earth's surface.

There's one more version of Kepler's Third Law that you may encounter. That version looks like this:

$$P^2 = \frac{4\pi^2 a^3}{GM}. \tag{2.17}$$

This version of Kepler's Third Law uses standard (SI) units: seconds for P, meters for a, and kilograms for M, and G is the universal gravitational constant, as described in Section 2.1.1. As you'll see in one of the following exercises, this version gives the same results as Eq. 2.15 provided you use the appropriate units in each case.

Exercise 2.9. **The dwarf planet Pluto orbits the Sun approximately once every 248 years. What is the semi-major axis of Pluto's orbit?**

Exercise 2.10. **Jupiter's moon Europa orbits Jupiter with a period of about 3.55 Earth days, and the semi-major axis of Europa's orbit is about 671,000 miles. What is the mass of Jupiter?**

Exercise 2.11. **Show that Eqs. 2.13, 2.16 and 2.17 give the same answer for the period of an asteroid orbiting the Sun in an orbit with semi-major axis of 3.2 AU.**

2.4 Chapter problems

2.1 Calculate the force of gravity between Jupiter and the Sun and the ratio of the accelerations of Jupiter and the Sun due to their mutual gravity.

2.2 (a) How does your result for the force of Earth's gravity on a 50-kg person (Exercise 2.2) compare to the result of the example done in Section 2.1.3 for a 100-kg person? Explain why these results make sense. (b) Does a higher-mass planet always have a larger force of surface gravity? Explain why or why not.

2.3 When the Sun expands into a red giant near the end of its life, it will have 100 times larger radius but roughly the same mass as it has now. How will the gravity at the Sun's new "surface" compare to its current surface gravity?

2.4 How does your weight at the top of Mount Everest (8,848 m above sea level) compare to your weight at the bottom of the Marianas Trench (10,994 m below sea level)?

2.5 Calculate the acceleration of the Earth due to the Moon's gravity and the acceleration of the Moon due to the Earth's gravity.

2.6 A spacecraft is towing a spacebarge which has no engines. Suddenly, the spacecraft runs out of fuel and its engines shut down. Both craft are very far away from any planet or star, so there are no external forces acting upon them. Describe the motion of each craft after the spacecraft's engines shut down. Would it matter if the tether were cut after the engines turn off?

2.7 Consider the following ellipse:

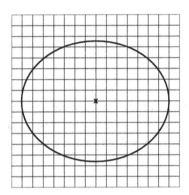

(a) Find the eccentricity of this ellipse.

(b) If this is a planet's orbit, find the location of the star it is orbiting.

(c) If each division of the grid is 1 AU and the star has mass of 1 solar mass, what is the period of the planet's orbit?

2.8 Imagine a planet with an orbit for which the aphelion is 15% larger than the perihelion. What is the eccentricity of the orbit?

2.9 How far from the center of the Earth is the International Space Station if its orbital period is 92 minutes?

2.10 Exoplanet 55 Cancri d orbits its parent star in an orbit with semi-major axis of 5.76 AU and period of 14.3 Earth years. What is the mass of the star?

2.11 Proponents of *astrology* (which we hope you never confuse with astronomy) assert that the positions of the planets can somehow influence our lives. To evaluate whether the force of gravity from other planets could possibly be responsible for any influence, compare the force of gravity on a newborn baby from (a) the planet Jupiter and (b) the doctor who delivers the baby. Make reasonable estimates for the mass of a doctor and his or her distance from a baby. For Jupiter's distance from the baby, use Jupiter's orbital semi-major axis.

3
Light

In astronomy, virtually all of the information that we can learn about the Universe comes from various forms of light.[1] Since planets, stars, and other objects in space are so far away and our ability to travel in space is rudimentary, we must glean as much information as possible from their light. Therefore, it behooves you to understand how light works and what kind of information it carries.

A great deal of the information in light from astronomical objects can be derived from the *spectrum* of that light. You can read about astronomical spectra in the first section of this chapter, and the later sections discuss some of the techniques astronomers use to interpret spectra.

3.1 Light and spectrum fundamentals

The most fundamental property that distinguishes one type of light from another is its color. This section introduces the concept of a spectrum as a graphical presentation of the brightness of different colors in light, and you'll learn how to translate between various quantitative properties associated with the color of light. Light behaves both as waves and particles, and you'll see how the properties related to color can be used to describe both the electromagnetic-wave and the photon-particle aspects of light. If you'd like to understand why light is called an electromagnetic wave and exactly what's doing the waving in light, you can find additional resources about the nature of light on the book's website.

---

[1] In astronomy, the word "light" is often used to refer to any type of electromagnetic radiation, which includes radio waves, microwaves, infrared raditaion, visible light (also called optical light), ultraviolet radiation, X-rays, and gamma rays.

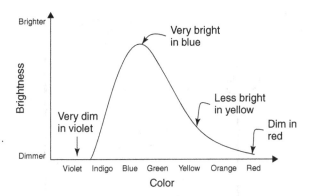

Figure 3.1 A spectrum of visible light.

3.1.1 Spectra

All forms of light are collectively referred to as the "electromagnetic spectrum." Used in this way, the word "spectrum" (plural "spectra") refers to the entire range of types of light found in nature, in the same way that "spectrum of political thought" refers to a wide range of opinions. In science, however, the term spectrum is also used in a slightly different way that has a special meaning. The spectrum of an object is a graphical representation of the amount of each color of light present in the object's radiation. The horizontal axis of a spectrum represents the color of light (or one of its proxies such as wavelength, frequency, or energy, which you can read about later in this section) and the vertical axis represents the amount of light (which may be called the brightness, intensity, or energy flux). An example of a spectrum is shown in Figure 3.1.

As you can see in this figure, the height of the line in the vertical direction indicates the brightness of each color of the light.

When we detect light using our eyes (or a telescope), all the colors of the light are mixed together. In order to produce a spectrum, the colors need to be separated using a prism or diffraction grating, which then allows us to determine how much of each color is present (under certain conditions, raindrops can act as tiny prisms to separate the colors in sunlight, which is how rainbows form). The brightness of each color can then be represented visually on a graph such as Figure 3.1.

It may help you to understand the information on a spectral graph by considering a spectrum not of light but of sound waves, with which you may be more familiar. Think about the sound waves produced by pressing the keys of a piano. Instead of making a graph of the waves themselves, imagine making

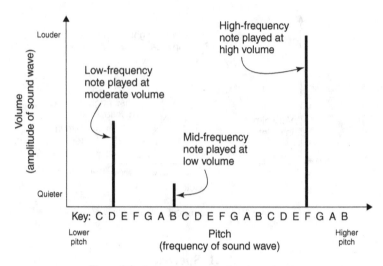

Figure 3.2 A spectrum graph of a piano chord.

a kind of bar graph to represent all of the sounds you hear simultaneously, in which the left–right position of the bar represents the frequency or pitch of the sound (how high or low the tone), and the height of the bar represents the intensity or volume of the sound (how quiet or loud it is). Such a graph is shown in Figure 3.2.

To make the sound represented by this spectrum, you'd have to simultaneously press the low-D, middle-B and high-F keys. But if you pressed all three keys equally hard, all three bars in the spectrum would be the same height. So for this chord, you'd have to simultaneously press the low-D key with moderate force, the middle-B key very lightly, and the high-F key very strongly. We emphasize *simultaneously* because it's a common mistake for students to think that a spectrum shows frequencies that are occurring at different times, which is not the case. A spectrum is a "snapshot in time" of the waves that are being emitted (or received) at the same time.

Exercise 3.1. Draw one frequency spectrum of the chord C, E, G for the case in which all three notes are played with equal (high) volume, and another spectrum in which all the notes in an entire octave (from C to C) are played with equal (low) volume.

To make the transition from a spectrum of sound waves to a spectrum of light waves, just remember that color is to light as pitch is to sound – these are quantities related to the *frequency* of the wave, which will be discussed in much more detail later in this section. Likewise, brightness is to light as volume is to sound – these are quantities related to the intensity, or *amplitude* of the wave. So, on a spectral graph for light, the horizontal axis represents one of the quantities related to color, and the vertical axis represents the brightness of each color. To determine whether a graph is a spectrum, just look at the horizontal axis to see if it represents some measure of color (such as frequency, wavelength, or energy). If so, it's a good bet that the graph is a spectrum. To make sure, look at the vertical axis to see if it represents some measure of the amount of radiation (which may be called intensity, energy flux, or irradiance).

An example of an astronomical spectrum, in this case a spectrum of our Sun's light after passing through Earth's atmosphere, is shown in Figure 3.3. The axes are labeled with several of the many possible variations with which you might find a spectrum labeled. From this spectrum, you can tell how much light of each color is present. For example, the greatest amount of light is in the green (G), and there is less red (R) light than orange (O) or yellow (Y).

Notice that in this figure, the same spectrum is presented in two different ways: in the left half of the figure, the spectrum is drawn with wavelength increasing to the right (which means that frequency and energy are increasing to the left, for reasons discussed below). In the right half of this figure, the same spectrum is drawn with frequency and energy increasing to the right (which means that wavelength is increasing to the left). The same information is presented in both graphs.

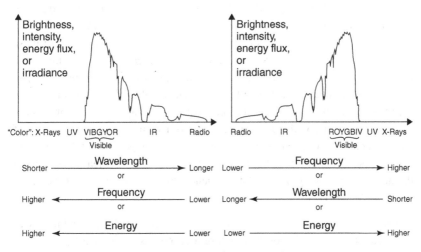

Figure 3.3 Examples of spectrum graphs.

Exercise 3.2. In the spectrum shown in Figure 3.3, rank the relative brightness of visible light (which includes ROYGBIV – red, orange, yellow, green, blue, indigo, and violet light), ultraviolet (UV), infrared (IR), and radio waves.

3.1.2 Relationship of wavelength, frequency, and energy

To understand why frequency, wavelength, and energy all relate to the color of light, it's necessary to understand the meaning of each of these terms. Here's a quick summary.

Wavelength (λ) Wavelength is the distance between adjacent peaks of the wave (or, equivalently, between adjacent valleys). It is not the total length of the wave, since a wave may consist of many oscillations, but rather the physical distance between neighboring oscillations in the electric and magnetic fields that make up a light wave. Since wavelength represents a distance per cycle (where "cycle" refers to one complete oscillation of the wave), the complete units of wavelength are meters per cycle (m/cycle). However, the common practice is to omit the "per cycle" portion of these units, so the standard units of wavelength are meters (m).

Frequency (f) Many students struggle to grasp the exact meaning of a wave's frequency. Your astronomy book probably says something like "Frequency is a measure of the rate at which wave peaks pass a fixed point in space," and that's a fine definition. But to develop an intuitive understanding of frequency, just ask yourself how frequently different events happen in your everyday life: how frequently do you eat lunch, or call your mother, or blink? Perhaps once per day, twice per week, and ten times per minute, respectively. Each of these answers is an expression of a frequency, which you can tell by looking at the units. In each of these cases, all units fit the same pattern: number of events *per unit time*. One lunch/day, two calls/week, and ten blinks/minute all represent the rates at which events occur. Hence frequency always has units of number of events per time, with time usually represented in seconds. In the case of waves (including light), frequency has units of cycles per second, but the term "cycles" is not included explicitly in common practice. This leaves no units in the numerator, making the unit of frequency "per second" or 1/s. This unit is called the hertz (Hz).

Energy (E) Light waves are made up of tiny bundles of electric and magnetic fields called "photons." You can picture a photon as a tiny wave packet that behaves like an individual "particle" of light – the smallest unit of light that can exist. Photons and waves are two different but equally valid ways to think about and describe light, and most astronomy textbooks use both, choosing whichever description is most convenient in various circumstances.

The energy of a photon is directly proportional to the frequency of the waves that make up the photon so $E \propto f$. This means that a photon of blue light has more energy than a photon of red light, since blue light has higher frequency than red light. A macroscopic analogy may help you remember this: imagine walking at a constant speed (representing the wave speed) while waving your arm up and down to represent the oscillations of a wave. If you wave your arm more frequently, you'll use more energy in going a given distance even though your walking speed was the same in both cases. You can read more about the energy in light waves a bit later in this section.

Many problems in astronomy require you to convert between the frequency and the wavelength of a light wave. This is made easier by the fact that all light waves (or, equivalently, photons) traveling through the vacuum of space have the same speed, irrespective of their wavelength, frequency, or energy. Surprisingly, that speed does not depend on the motion of the light source or the observer. The speed of light in a vacuum is 3×10^8 m/s and is represented by the letter c.

Given the constancy of the speed of light, the relationship between wavelength and frequency can be understood by picturing two waves, one with long wavelength and the other with short wavelength, propagating at the same speed. Now ask yourself how frequently the peaks of each of these waves will travel past a given point. Clearly, the peaks of the shorter-wavelength wave will travel past more frequently than the peaks of the longer-wavelength wave, because the distance between the peaks is greater for the long-wavelength wave. So as long as the speed of the waves is the same, *longer wavelength* must correspond to *lower frequency*.

This analysis means that the only thing that determines the frequency of a light wave is its wavelength, and that wavelength and frequency are *inversely* proportional (that is, one is bigger if the other is smaller). Expressing this using the proportionality relationships discussed in Section 1.2, $\lambda \propto f^{-1}$, or $\lambda \propto 1/f$. Remember that "$\propto$" means "is proportional to" or "equals a constant times." In this case, the constant is the speed of light (c), so the equation relating wavelength and frequency is

$$\lambda = \frac{c}{f} \quad \text{or} \quad f = \frac{c}{\lambda} \quad \text{or} \quad \lambda f = c. \tag{3.1}$$

You can use this equation to calculate either the wavelength (λ) or frequency (f) of any light wave, given the other.

Example: The visible portion of the electromagnetic spectrum is centered near a wavelength of 500 nm (1 nm = 10^{-9} m). What is the frequency of these waves?

Choose the version of Eq. 3.1 solved for frequency and plug in 500×10^{-9} m for λ and 3×10^8 m/s for c:

$$
\begin{aligned}
f &= \frac{c}{\lambda} \\
&= \frac{3 \times 10^8 \text{ m/s}}{500 \times 10^{-9} \text{ m}} \\
&= \frac{3 \times 10^8}{5 \times 10^{-7}} \frac{1}{\text{s}} = \frac{3}{5} \times 10^{8+7} \text{ Hz} \\
&= 0.6 \times 10^{15} \text{ Hz} = 6 \times 10^{14} \text{ Hz}.
\end{aligned}
$$

Note that when the units of meters canceled, seconds retained its place in the denominator, so meters *per* second (m/s) became *per* second (1/s), or hertz (Hz). We emphasize the word "per" because many students drop the fraction bar that represents "per" in that step, so their units erroneously come out in seconds – which is not a unit of frequency.

Since the frequencies of electromagnetic waves are typically very large numbers in hertz (especially in the visible range and above), they are often preceded by metric prefixes. Thus you're very likely to run across frequencies expressed with units of kilohertz (kHz, or 1000 Hz), megahertz (MHz, or 10^6 Hz), gigahertz (GHz, or 10^9 Hz), and terahertz (THz, or 10^{12} Hz). Wavelengths, on the other hand, are often very small, so you will often find wavelengths expressed with smaller metric prefixes such as millimeters (mm, or 10^{-3} m), micrometers or "microns" (μm, or 10^{-6} m), and nanometers (nm, or 10^{-9} m). Though not a metric unit, angstroms (Å, or 10^{-10} m) are also commonly used to express wavelengths.

In addition to finding the wavelength of an electromagnetic wave for which you know the frequency and vice versa, you may also be asked to find the energy of the photons of that wave. As mentioned above, the energy of a photon is directly proportional to the photon's frequency, and the constant of proportionality is Planck's constant ($h = 6.626 \times 10^{-34}$ J s). The unit J stands for joules, the SI unit for energy. Thus the equation for converting between frequency and photon energy is

$$E = hf. \tag{3.2}$$

Example: Find the energy in a photon with λ = 500 nm.

To find the energy, you can plug the values for h and f (found in the previous example) into Eq. 3.2, remembering that the unit Hz is the same as s^{-1}:

$$E = h\,f = \left(6.626 \times 10^{-34}\,\text{J}\,\cancel{\text{s}}\right)\left(0.6 \times 10^{15}\,\frac{1}{\cancel{\text{s}}}\right) = 3.98 \times 10^{-19}\,\text{J}.$$

You might find yourself wondering "If energy is related to frequency, and frequency is related to wavelength, how is energy related to wavelength?" You can answer that question by referring back to the previous analogy of waving your arm, and remembering that a larger energy corresponded to a higher frequency – waving your arm up and down faster while walking at the same speed. The arm-waving analogy also illustrates the inverse relationship between wavelength and frequency, because the distance covered between successive waves of your arm – imagine dropping a pebble on the floor with each wave and measuring the distance between pebbles – decreases as your frequency of waving increases. So as you wave your arm more furiously (higher frequency and higher energy), the distance between pebble drops (wavelengths) decreases. Thus you can expect that energy and wavelength should be *inversely* proportional: $E \propto 1/\lambda$. Combining Eqs. 3.1 and 3.2 bears this out:

$$E = h\,f = h\left(\frac{c}{\lambda}\right) = \frac{h\,c}{\lambda}. \tag{3.3}$$

Since wavelength is in the denominator on the right side of this equation, longer wavelength corresponds to lower energy, and shorter wavelength corresponds to higher energy.

Exercise 3.3. Find the wavelength of an electromagnetic wave with frequency of 3.2 GHz.

Exercise 3.4. Find the frequency of a photon of red light with wavelength of 630 nm.

Exercise 3.5. Find the energy of the red photon in the previous exercise.

3.2 Radiation laws

When you stand outdoors on a clear evening and look up at the night sky, the vast majority of objects you see are visible for one reason: they're emitting light because they're hot.[2] The laws which describe the color and the amount of light given off by hot objects are called radiation laws, and most astronomy texts include a discussion of these laws.

The traditional way of referring to the light given off by a hot object is to use the contradictory-sounding term "blackbody radiation," and objects emitting this kind of light are called "blackbodies." Students are often surprised to learn that a blackbody appears almost any color *except* black, unless its temperature is absolute zero. This is relevant to astronomy because stars behave very much like blackbodies. So although a blackbody is generally not black, there is a good reason for the term blackbody. That reason is that a blackbody does not reflect any light that falls upon it (hence "black"). Instead, a blackbody absorbs all incoming radiation and all the energy of that radiation. This absorbed energy contributes to heating the object, which may have its own internal heat sources as well. But here's the really useful thing about a blackbody: it emits a very predictable spectrum of radiation with a characteristic shape that depends solely on its temperature. So to analyze the radiation emitted by a blackbody, you don't have to worry about the amount or type of incoming radiation at all. The blackbody could be getting its energy from incoming light, radio waves, X-rays, or almost entirely from its own internal energy sources, but all that matters is the total amount of that energy, which determines the temperature of the blackbody. And if you know the temperature of a blackbody, you know the exact shape of the spectrum of radiation that it emits.

You can see an example of that shape in the spectrum graph of an 8,000-kelvin blackbody shown in Figure 3.4.

Notice that in this spectrum wavelength is shown on the horizontal axis and increases to the right, and the amount of electromagnetic radiation is indicated on the vertical axis as "emitted flux." This shows how brightly the blackbody shines at each wavelength, and in this plot the brightness has been "normalized" to the peak value, which means that it has been scaled so that the peak of the spectrum has a y-value of 1, or 100%.

Since temperature is the key parameter that determines the color and amount of light from a blackbody, a far more revealing term for this type of light is "thermal radiation." In this book we use the terms thermal and blackbody radiation interchangeably.

As the temperature of a blackbody changes, both the *color* and the *amount* of thermal radiation emitted from that blackbody change. You can see this in the three thermal-radiation curves shown in Figure 3.5.

---

[2] A very few objects in the night sky – moons and planets, for example – are visible because they are reflecting light from a hot object – the Sun.

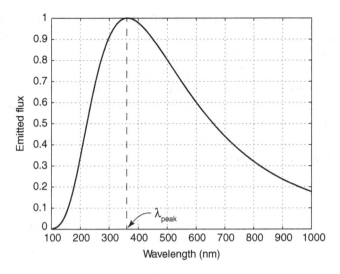

Figure 3.4 A blackbody spectrum.

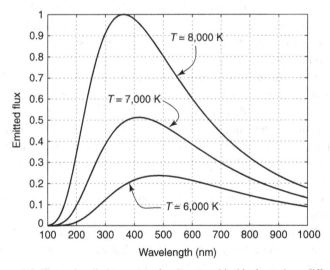

Figure 3.5 Thermal radiation curves for the same blackbody at three different temperatures.

These three curves show the energy flux of the thermal radiation emitted from the same blackbody at three different temperatures: 6,000 K, 7,000 K, and 8,000 K. Notice that increasing the temperature of this blackbody causes the wavelength at which the spectrum reaches its peak (called λ_{peak}) to shift toward shorter wavelengths (to the left in this figure). But you can clearly see another effect: the height of the peak (that is, the amount of radiation) increases significantly as temperature increases.

The shifting in a blackbody's spectrum is described by two radiation laws: Wien's Law and Stefan's Law. Wien's Law relates a blackbody's temperature to the color of its radiation, and Stefan's Law relates a blackbody's temperature to the amount of radiation the blackbody emits (specifically, to the energy flux emitted from each square meter of the blackbody's surface). Wien's Law and Stefan's Law are the subjects of the next two subsections.

3.2.1 Wien's Law

The equation that relates an object's temperature (T) to its color is Wien's Law, which can be written like this:

$$\lambda_{peak} = \frac{b}{T},\tag{3.4}$$

where the constant we're calling "b" has a value of $b = 0.0029\,\text{m K}$ in SI units.[3] In this equation, λ_{peak} represents the wavelength at which the spectrum peaks (that is, the wavelength at which the greatest amount of radiation is emitted) in meters, and T is the object's temperature in kelvins.

You can see that when you plug a temperature in kelvins (K) into Wien's Law, the units of K will cancel in the numerator and denominator, leaving units of meters, the standard units of wavelength. With this equation, you can calculate peak wavelength if you're given the temperature, but in astronomy Wien's Law is most often used in the other direction – the wavelength of the spectral peak is measured in order to determine the temperature of a celestial object. You should note that this equation works only when λ_{peak} has units of meters. So if you have (or want) a value for wavelength in units other than meters, you will need to do a unit conversion on the value of either the peak wavelength or the constant b in order to make their length units match.

It is very important for you to remember that λ_{peak} is the wavelength at which the spectrum peaks, not the height of the spectrum at the peak. That is, it is the (horizontal) x-value of the peak, not the (vertical) y-value. Thermal radiation spectra from perfect blackbodies always have the shape shown in Figure 3.4 with one unique peak. For calculating temperature using Wien's Law, all that matters is how far left or right the peak lies on the spectrum. In fact, this law is often referred to as Wien's "displacement" law, emphasizing

---

[3] Some texts express b in other units, such as $0.29\,\text{cm K}$ or $2.9\,\text{mm K}$.

the fact that the peak of the spectrum gets displaced (shifted) to the left or right as temperature varies.

So does higher temperature shift the spectrum of thermal radiation to the left or to the right? The answer to that question depends on exactly what is plotted on the horizontal axis. Many astronomy texts show wavelength increasing (and therefore frequency decreasing) to the right, but some show frequency increasing (and wavelength decreasing) to the right. Whichever direction your book plots increasing wavelength, you can be certain of one thing: peak wavelength and temperature are *inversely* proportional: $\lambda_{peak} \propto 1/T$. This means that as temperature gets bigger, the peak wavelength gets smaller, so the spectrum shifts in the direction of decreasing wavelength (which is the same as the direction of increasing frequency).

Wien's Law means that for blackbodies emitting light in the visible range, hotter means bluer and cooler means redder (since blue light has shorter wavelength than red light). Knowing this allows you to make a fundamental astrophysical measurement using only your eyes when you look up at the night sky. Since stars are reasonably good blackbodies, those that appear somewhat red must have a lower temperature than those that appear blue. And our Sun, which has λ_{peak} in the yellow–green range, is not among the hottest or the coolest stars, residing comfortably in the middle range of stellar temperatures.

There is a good physical reason for the inverse relationship between temperature and wavelength, which you can understand if you think about energy. A hotter object has more internal energy, so on average it will radiate higher-energy photons. Since all thermal radiation curves have the same shape, if the average of the spectrum shifts toward higher energy, the peak must shift in the same direction. And, as Eq. 3.3 tells you, energy is inversely proportional to wavelength, so a spectrum shifted to higher energy must be shifted toward shorter wavelength (and higher frequency).

Here's an example of using Wien's Law (Eq. 3.4) to find the peak wavelength of the spectrum of a blackbody of known temperature:

Example: At what wavelength does the spectrum of the thermal radiation from a human being with body temperature of 310 K peak?

Since the temperature is already given in kelvins, you can plug $T = 310$ K directly in to Eq. 3.4:

$$\lambda_{peak} = \frac{0.0029 \text{ m} \cdot \cancel{K}}{310 \, \cancel{K}} = \frac{2.9 \times 10^{-3}}{3.1 \times 10^2}\text{m} = \frac{2.9}{3.1} \times 10^{-5} \text{ m} = 0.94 \times 10^{-5} \text{ m}.$$

This result shows you that humans give off most of their thermal radiation at about 9.4 µm, which is in the infrared portion of the electromagnetic spectrum.

Does this answer make sense? In other words, are you glowing in infrared light right now? Yes, you are. But human eyes aren't sensitive to infrared light, which is why law-enforcement officials use infrared goggles to spot fleeing suspects in the dark.

Here's an example that goes the other way – using λ_{peak} to find the temperature of a blackbody:

Example: What is the temperature of the photosphere (light-emitting layer) of our Sun, whose spectrum peaks in the center of the visible part of the electromagnetic spectrum at about 500 nm?

The first step is to solve Eq. 3.4 for T:

$$\lambda_{peak} = \frac{b}{T} \rightarrow \lambda_{peak} \times \left(\frac{T}{\lambda_{peak}}\right) = \frac{b}{T} \times \left(\frac{T}{\lambda_{peak}}\right)$$

$$T = \frac{b}{\lambda_{peak}} = \frac{0.0029 \text{ m K}}{\lambda_{peak}}, \tag{3.5}$$

and then plug in λ_{peak}, using the conversion factor between nanometers and meters:

$$T = \frac{0.0029 \text{ m} \cdot \text{K}}{500 \text{ nm}} \left(\frac{1 \text{ nm}}{10^{-9} \text{ m}}\right) = \frac{2.9 \times 10^{-3}}{5 \times 10^2 \times 10^{-9}} \text{ K} = \frac{2.9}{5} \times 10^4 \text{ K}$$

$$= 5,800 \text{ K}.$$

This is the temperature of the outer layer of the Sun, but the temperature of the Sun's core is much higher – around 15 million kelvins.

3.2.2 Stefan's Law

Stefan's Law allows you to calculate the power of the radiation produced by each square meter of that blackbody's emitting surface as a function of temperature. That power per area is sometimes called energy flux (EF), and Stefan's law is usually written like this:

$$EF = \sigma T^4, \tag{3.6}$$

where EF has units of watts per square meter ($\frac{W}{m^2}$), σ is called "Stefan's constant" ($\sigma = 5.67 \times 10^{-8} \frac{W}{m^2 K^4}$), and T represents the temperature of the blackbody in kelvins.[4] Some textbooks refer to Stefan's Law as the Stefan–Boltzman Law, and some include in this equation a factor called the

---

[4] Here is an easy way to remember Stefan's constant: since the coefficient is 5.67, and the exponent is (negative) 8, just remember counting 5, 6, 7, 8.

"emissivity" of the object – but the emissivity of an ideal blackbody is one, and many astronomical objects (including stars) are very similar to ideal blackbodies in this respect.

A quick look at this equation tells you two things. Since temperature is in the numerator, hotter objects emit more radiation (they have more energy, so they pump out more photons). And since temperature is raised to the fourth power, small changes in temperature produce large changes in the radiated power. For example, doubling the temperature doesn't just double the energy flux, it increases the energy flux by a factor of $2^4 = 16$.

For some problems, you may be interested not in the power given off by each square meter of a blackbody, but in the total power radiated by the whole object. In astronomy, total power is called "luminosity," which has units of watts (W) and is a measure of the total energy radiated by an object per unit time (since watts are equivalent to joules per second).

The luminosity of an object depends not only on the amount of power emitted per unit area, but also on the size of the object – specifically, on the surface area (SA) from which radiation is emitted:

$$L = SA \times EF = (SA)\sigma T^4. \tag{3.7}$$

For the simplest and most symmetric case of a sphere, which is relevant because many objects in astronomy are approximately spherical in shape, the surface area is given by $SA = 4\pi R^2$. Combining the expressions for luminosity and surface area with Stefan's Law (Eq. 3.6), the equation for the luminosity of a spherical blackbody of radius R and temperature T is

$$L = 4\pi R^2 \sigma T^4, \tag{3.8}$$

which you can see in expanded form in Figure 3.6.

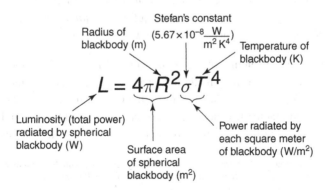

Figure 3.6 The luminosity of a spherical blackbody.

If your astronomy text uses proportionality relationships, you may find a version of this equation that looks like this:

$$L \propto R^2 T^4. \tag{3.9}$$

This proportionality relationship is very handy for seeing the dependence of luminosity on radius and temperature, and it is ideally suited for use in problems that you can solve using the ratio method.

Example: The planets Earth and Venus are approximately the same size, but Venus is significantly hotter ($T_{Earth} \approx 290$ K and $T_{Venus} \approx 700$ K). How much more luminous is Venus?[5]

This problem provides an excellent example of the power and simplicity of the ratio method, since you're given the relative rather than the absolute values of their radii (the problem states that they are "approximately the same size"). So you can start by writing Eq. 3.9 for both Venus and Earth and then forming a ratio by dividing one equation by the other. Since you know that $R_E \approx R_V$, you can treat them as equal and let them cancel:

$$\frac{L_V \propto \cancel{R_V^2} T_V^4}{L_E \propto \cancel{R_E^2} T_E^4} \rightarrow \frac{L_V}{L_E} = \frac{T_V^4}{T_E^4} = \left(\frac{T_V}{T_E}\right)^4 = \left(\frac{700\,\cancel{K}}{290\,\cancel{K}}\right)^4 \rightarrow \frac{L_V}{L_E} \approx 33.9.$$

This result, $L_V \approx 33.9\, L_E$, tells you that the luminosity of Venus is about 34 times that of Earth. Its temperature is only 2.4 times greater, but since $L \propto T^4$ (with equal radii), this changes the luminosity by $(2.4)^4 = 33.9$ times.

Now consider the question of how the luminosities compare between two objects of the same temperature but different size.

Example: Imagine two stars of the same temperature: a small spectral type M star, and a red giant.[6] *If the red giant is 1,000 times larger in radius (as the name implies), how do the two stars' luminosities compare?*

Since you are not given any of the actual values for T and R for these stars, this problem calls for the ratio method. You are told that the stars' temperatures are equal, so $T_M = T_{RG}$. You can substitute either of these variables for the other, which will facilitate canceling, or you can just cancel them straightaway as shown below. Furthermore, the radius of the red giant is 1,000 times larger

---

[5] Remember that planets like Earth and Venus "shine" by reflecting sunlight, but they also emit their own thermal radiation (mostly in the infrared, as you can tell by plugging their temperatures into Wien's Law).

[6] You can read about spectral types and red giants in most comprehensive astronomy textbooks.

than the small M star, so $R_{RG} = 10^3 R_M$. Thus where you see R_{RG} in the equation, you can substitute $10^3 R_M$:

$$\frac{L_M \propto R_M^2 T_M^4}{L_{RG} \propto R_{RG}^2 T_{RG}^4} \rightarrow \frac{L_M}{L_{RG}} = \left(\frac{R_M}{R_{RG}}\right)^2 = \left(\frac{R_M}{10^3 R_M}\right)^2.$$

Many students make a critical mistake in problems like this by forgetting to square the factor of 10^3. It's very important for you to realize that after you substitute $10^3 R_M$ for R_{RG}, the R_Ms cancel, but the factor of 10^3 and the power (^{2}) do not cancel. Introducing parentheses and placing any powers outside the parentheses before substituting will help you avoid inadvertently canceling factors and powers that you still need. After canceling and simplifying, the remaining terms are

$$\frac{L_M}{L_{RG}} = \frac{1}{10^6}.$$

Equivalently,

$$L_M = \frac{1}{10^6} L_{RG}, \text{ or } L_{RG} = 10^6 L_M.$$

So in this case the red giant is a million times more luminous than the small M star. Since their temperatures are the same, this difference is due entirely to the larger size of the red giant. Even though the radius is only 10^3 times bigger, since $L \propto R^2$ (with T constant), the luminosity is $(10^3)^2 = 10^6$ times higher. It is this high luminosity that makes red giants easy to find, even though they are not nearly as common as small M stars.

The previous examples were well suited to the use of the ratio method, but you may also encounter problems in which you're asked to calculate actual values for luminosity by plugging in real values with units for the variables and keeping all the constants. Here's an example of such a problem using the absolute method.

Example: What is the luminosity of the human body? That is, how many watts does the typical living human being radiate? For this problem, you can assume a body temperature of $T = 310$ K and skin surface area of 1.5 m².

Since you're given the temperature and surface area of the object and asked to find the luminosity, Eq. 3.7 is the relation to start with:

$$L = SA \times EF = (SA)\sigma T^4. \tag{3.7}$$

Plugging in the values given in the problem statement gives

$$L_{human} = (SA_{human}) \sigma T_{human}^4 \simeq 1.5 \, m^2 \left(5.67 \times 10^{-8} \frac{W}{m^2 \cdot K^4}\right) (310 \, K)^4.$$

Notice that the temperature units of K^4 and $(K)^4$ cancel because the power of 4 outside the parentheses applies to the units. But you must apply this fourth power to the number (310) as well. Multiplying gives

$$L_{human} \approx 785 \text{ W}.$$

This result says that your body is radiating a significant amount of power. However, it assumes perfect radiation efficiency (emissivity $= 1$) from human skin, and it represents the power (energy per time) you are radiating *into* your environment without considering the power you are also absorbing *from* your environment. Allowing for those effects, your *net* rate of energy transfer to your environment by radiation is typically closer to 100 W. Most of this energy is emitted as infrared light, which explains why you don't see other people visibly glowing – but this energy does cause a crowded room full of people to tend to warm up.

Example: How does the luminosity compare between two objects whose temperature and size are both different? Consider the Earth and Sun: the Sun's photosphere temperature is 5,800 K and the Earth's temperature is 290 K, and the Sun is 100 times larger in radius than the Earth. For this example, you can treat both objects as ideal blackbodies, which is a good assumption for the Sun, but less so for the Earth.

You are given values for the two temperatures, but only the ratio of the radii. While you could look up the radii and use the absolute method, this is not necessary. You can instead express one radius in terms of the other, $R_S = 100 R_E$, and then make a substitution:

$$\frac{L_S \propto R_S^2 T_S^4}{L_E \propto R_E^2 T_E^4} \rightarrow \frac{L_S}{L_E} = \left(\frac{R_S}{R_E}\right)^2 \left(\frac{T_S}{T_E}\right)^4 = \left(\frac{100 R_E}{R_E}\right)^2 \left(\frac{5,800 \text{ K}}{290 \text{ K}}\right)^4.$$

Notice that with the ratio approach, the variable representing the radius of Earth (R_E) canceled, leaving only the numerical factor from the ratio of the radii. Also, the units of temperature cancelled, though their numerical values did not. Simplifying the remaining numbers gives

$$\frac{L_S}{L_E} = (100)^2 \left(\frac{5,800}{290}\right)^4 = (10^2)^2 (20)^4 = 10^4 \times 1.6 \times 10^5 = 1.6 \times 10^9.$$

This result tells you that our star (the Sun) is over 1 billion times more luminous than our planet (the Earth). Does this answer make sense? Yes, it does. Stars are typically millions to billions of times brighter than their planets, which is one reason why planets orbiting around distant stars are exceedingly difficult to find – their light is completely swamped by the light of their stars.

Of course, planets also "shine" by reflecting light from the star they are orbiting, but that reflected light is a negligible fraction of the light produced by the star.

You could also have worked this problem by expressing one temperature as a multiple of the other (in this case, $T_S = 20T_E$) and then substituting as you did for radius:

$$\frac{L_S}{L_E} = \left(\frac{R_S}{R_E}\right)^2 \left(\frac{T_S}{T_E}\right)^4 = \left(\frac{100\cancel{R_E}}{\cancel{R_E}}\right)^2 \left(\frac{20\cancel{T_E}}{\cancel{T_E}}\right)^4 = (10^2)^2(20)^4 = 1.6 \times 10^9.$$

For more practice with problems involving Stefan's Law, using both the ratio method and the absolute method, see the problems at the end of the chapter and the on-line solutions.

3.2.3 Applying the radiation laws

To understand why the radiation laws are so powerful in astronomy, you have to think about the difference between the energy flux emitted by a distant blackbody (such as a star) and the energy flux received here on Earth. The situation is illustrated in Figure 3.7.

As shown in the figure, the energy flux at the star refers to the amount of radiation given off by each square meter of the star's surface, which you know from Stefan's Law to be $EF = \sigma T^4$. But this is clearly not the energy flux *received at Earth* (called the apparent brightness of the star), because the star light spreads out as it travels. The amount reaching Earth depends on the total power given off by the star (the luminosity of the star) and on the distance from

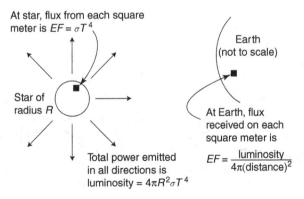

Figure 3.7 Energy flux at star and at Earth.

the star to the Earth. You can read about luminosity and apparent brightness in Section 5.2, but at this point the important concept is that the energy flux at the star depends only on the temperature of the star, but the energy flux received at Earth also depends on the size of the star and the star's distance from Earth.

The difference between the energy flux emitted by a blackbody and energy flux received by an observer can be seen by comparing the blackbody curves shown in the left and right portions of Figure 3.8.

The curves labeled 1 through 5 in the left portion of this figure represent the energy flux emitted by five blackbodies of different temperatures. Since the spectrum of blackbody 1 reaches a peak at a shorter wavelength than the spectra of blackbodies 2 through 5, Wien's Law tells you that blackbody 1 must have a higher temperature than the others. And since the emitted flux (the number of watts of power radiated by each square meter of the blackbody's surface) depends only on the temperature of the blackbody, you can also use Stefan's Law to determine that blackbody 1 is hotter than the others, since its emitted energy flux is higher at all wavelengths.

Now consider the curves in the graph on the right portion of Figure 3.8. These curves represent the energy flux received by an observer on Earth from the five blackbodies whose emitted flux is shown in the left portion of the figure. Here's the critical point: since the received flux depends on the distance of the blackbody from Earth and on the total power radiated by the blackbody (which depends on the size of the object), the height of these curves cannot be used to determine the temperature of the blackbodies 1 through 5. So Stefan's Law is not helpful for determining the temperature of these objects.

Fortunately, even if the received energy flux is all we have, Wien's Law can be still be used to find the temperature of the blackbodies. That's because

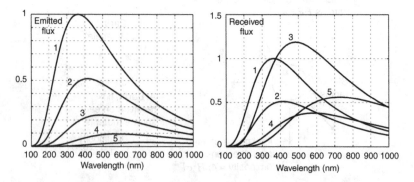

Figure 3.8 Emitted and received energy flux from five blackbodies.

Wien's Law requires only that we know the wavelength at which the spectrum reaches its peak. So for Wien's Law, the left–right position of the peak is all that's needed to determine the temperature – the height of the peak is not relevant. And although the different heights of the five curves makes it somewhat difficult to trace each one, if you look carefully you can see that the curve for blackbody 2 reaches its peak to the right of the peak for blackbody 1. Since wavelength increases to the right in this graph, that means that λ_{peak} for blackbody 2 is greater than λ_{peak} for blackbody 1, and Wien's Law tells you that λ_{peak} is inversely proportional to temperature. Hence blackbody 2 has a lower temperature than blackbody 1.

The same analysis applied to the curves for the other blackbodies indicates that blackbody 3 is cooler than blackbody 2, blackbody 4 has even lower temperature, and blackbody 5 is the coolest of the lot.

So if blackbody 3 has lower temperature than blackbodies 1 and 2, why is its curve higher? One possible reason is that blackbody 3 is bigger than the others – if it has more surface area, than the total power it radiates into space is greater, even though it gives off fewer watts per square meter. Another possible reason is that blackbody 3 may be closer than the others, and since these are graphs of received energy flux, the height of the curve depends on the distance to the object. But don't be fooled by the height of blackbody 3's curve – its peak occurs at a longer wavelength than the peaks of blackbodies 1 and 2, so blackbody 3 must be cooler than those two.

Exercise 3.6. Rank by temperature the objects that produce the received energy flux curves V, W, X, Y, and Z shown in Figure 3.9.

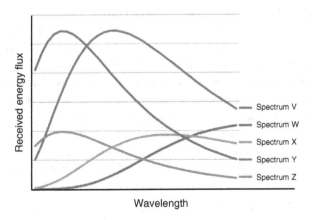

Figure 3.9 Received energy flux for five different blackbodies.

Exercise 3.7. Would your ranking of the objects by temperature change if the horizontal axis of the spectrum showed frequency instead of wavelength increasing to the right? What if the horizontal axis showed energy instead of wavelength? Explain your reasoning.

3.3 Doppler shift

One of the most powerful tools in an astronomer's toolkit is the ability to measure the speed with which an object is moving toward or away from the Earth. This tool is called "Doppler shift" and it works for any object that's emitting or reflecting electromagnetic waves.

3.3.1 The Doppler effect

You're probably already familiar with the Doppler effect, which you may have experienced as the changing pitch of a siren as its source moves past you. Although sound waves and light are very different, the Doppler phenomenon is at work in both cases: when the source of the waves is approaching you, the wave appears compressed and the wave's crests and troughs arrive more frequently. So in this case you measure a shorter wavelength and a higher frequency than the "true" values.[7] Likewise, if the source of the waves is moving away from you, the waves appear stretched out and the wave crests and troughs arrive less frequently than if the source were stationary. So for a receding source you measure a longer wavelength and a lower frequency than the true values.

Since blue light has shorter wavelength and higher frequency than red light, the light from approaching objects is said to be "blueshifted," and the light from receding objects is said to be "redshifted." In astronomy, this terminology is used even if the waves are not in the visible portion of the spectrum – a shift toward longer wavelengths is called a redshift and a shift toward shorter wavelengths is called a blueshift even if the waves are radio waves or X-rays. The Doppler effect also occurs when it's you (the observer) that's moving rather than the source – if you're moving toward the source, its light will appear blueshifted and if you're moving away from the source, its light will appear redshifted. All that matters is the *relative* speed between you and the source.

---

[7] The "true" values of wavelength and frequency are the values that are sent out by the source.

3.3.2 The Doppler equation

The equation relating the "apparent wavelength" (that is, the wavelength measured by the observer, λ_{app}) to the "true wavelength" (the wavelength given off by the source, λ_{true}) depends on two speeds: the speed of the source relative to the observer and the speed of the wave. For light, the speed of the wave in empty space is always c, and if the recession speed (defined as the speed with which an object is moving *away* from the observer) is v_{rec}, then the Doppler-shift equation may be written as

$$\frac{\lambda_{app}}{\lambda_{true}} = 1 + \frac{v_{rec}}{c}. \tag{3.10}$$

In this equation, λ_{app} and λ_{true} can be expressed in any units of length (including meters, km, μm, or nm) as long as both of these wavelengths are in the *same* units. Likewise, v_{rec} and c can be expressed in any units of speed (such as meters per second, kilometers per hour, or miles per hour) as long as they have the *same* units. Since v_{rec} is the speed of recession its value is positive if the source and observer are getting farther apart (receding) and negative if they're getting closer together (closing).

Of course, most things in the Universe are moving much more slowly than the speed of light, so the ratio v_{rec}/c is almost zero, which makes the right side of Eq. 3.10 very close to 1. And if the right side is about 1, then λ_{app} is almost identical to λ_{true}, and you have to make very precise measurements (and keep lots of decimal places in your calculations) to see any difference at all.[8] You can see that in the following example.

Example: A star is moving away from the Earth at a speed of 300 km/s. If that star is emitting radiation at a wavelength of 530 nm, what wavelength would be measured by an observer on Earth?

Since you're given the emitted wavelength λ_{true} and the speed of recession (v_{rec}) and you're trying to find the apparent wavelength λ_{app}, you can plug your values directly into the Doppler-shift equation (Eq. 3.10). But, as always, it's a good idea to begin by rearranging the equation to solve for the quantity you're trying to find (λ_{app} in this case) before plugging in any numbers:

$$\frac{\lambda_{app}}{\lambda_{true}} = 1 + \frac{v_{rec}}{c},$$

---

[8] Doppler-shift questions are among the few types of problems in introductory astronomy in which you should keep more than two or three decimal places. In most Doppler problems, it's advisable to keep six or more decimal places.

$$\lambda_{app} = \lambda_{true} \left(1 + \frac{v_{rec}}{c} \right),$$

$$\lambda_{app} = (530 \text{ nm}) \left(1 + \frac{300 \text{ km/s}}{3 \times 10^5 \text{ km/s}} \right)$$

(note that you must use the same units for c and v_{rec}). Thus

$$\lambda_{app} = (530 \text{ nm})(1 + 0.001) = 530.53 \text{ nm},$$

a change in wavelength of only 0.53 nm, which is 0.1% of the true wavelength.

Can such a small shift really be detected? After all, 530 nm is in the green portion of the spectrum, and so is 530.53 nm. So the "redshift" in this case does not change the color of the light from green to red. Fortunately, astronomers know the exact wavelength of many spectral lines with extreme precision, and tiny shifts in those spectral lines have been used to measure Doppler shifts from objects moving at speeds of less than 1 m/s relative to the Earth.

Exercise 3.8. If a spectral line from a source moving away from the Earth at a speed of 25 km/s appears to have a wavelength of 750 nm, what is the true wavelength of that line?

Exercise 3.9. If a spacecraft transmits a microwave signal with a wavelength of 1.250 cm and the wavelength of the signal received on Earth is 1.251 cm, how fast is the spacecraft moving away from the Earth?

3.3.3 Alternative forms of the Doppler equation

Another useful form of the Doppler-shift equation can be obtained by the following re-arrangement of Eq. 3.10:

$$\frac{\lambda_{app}}{\lambda_{true}} = 1 + \frac{v_{rec}}{c},$$

$$\lambda_{app} = \lambda_{true} \left(1 + \frac{v_{rec}}{c} \right) = \lambda_{true} + \lambda_{true} \left(\frac{v_{rec}}{c} \right),$$

$$\lambda_{app} - \lambda_{true} = \lambda_{true} \left(\frac{v_{rec}}{c} \right),$$

$$\frac{\lambda_{app} - \lambda_{true}}{\lambda_{true}} = \frac{v_{rec}}{c}.$$

Designating the change in wavelength ($\lambda_{app} - \lambda_{true}$) as $\Delta\lambda$, this becomes

$$\frac{\Delta\lambda}{\lambda_{true}} = \frac{v_{rec}}{c}, \qquad (3.11)$$

which tells you that the fractional change in wavelength is equal to the fraction of the speed of light at which the object is receding. For the values of the previous example, here's how that works:

$$\frac{\Delta\lambda}{\lambda_{true}} = \frac{v_{rec}}{c},$$

$$\frac{\Delta\lambda}{530 \text{ nm}} = \frac{300 \text{ km/s}}{3 \times 10^5 \text{ km/s}},$$

$$\Delta\lambda = 530 \text{ nm} \left(\frac{300 \text{ km/s}}{3 \times 10^5 \text{ km/s}}\right)$$

$$\Delta\lambda = 0.53 \text{ nm}.$$

Both Eqs. 3.10 and 3.11 can be used when the wave source and the observer are getting closer rather than receding, provided that you realize that the speed of recession (v_{rec}) is *negative* when the objects are getting closer. So if the star in the previous example had been approaching the Earth, you would simply have to set v_{rec} to -300 km/s in Eqs. 3.10 and 3.11. In that case, λ_{app} comes out shorter than λ_{true}, and $\Delta\lambda$ comes out negative, as you can see in the exercises at the end of this section.

In your astronomy studies, you may also encounter a version of the Doppler-shift equation that involves frequencies instead of wavelengths. You can easily convert from wavelength (λ) to frequency (f) and vice versa by applying Eq. 3.1 to the approaching and true values of wavelength and frequency:

$$\lambda_{app} f_{app} = c$$

and

$$\lambda_{true} f_{true} = c,$$

and then substituting $\lambda_{app} = c/f_{app}$ and $\lambda_{true} = c/f_{true}$ into the Doppler equation:

$$\frac{\lambda_{app}}{\lambda_{true}} = \frac{\frac{c}{f_{app}}}{\frac{c}{f_{true}}} = 1 + \frac{v_{rec}}{c},$$

$$\left(\frac{\cancel{c}}{f_{app}}\right)\left(\frac{f_{true}}{\cancel{c}}\right) = 1 + \frac{v_{rec}}{c},$$

$$\frac{f_{true}}{f_{app}} = 1 + \frac{v_{rec}}{c}. \tag{3.12}$$

Notice that since frequency and wavelength are inversely proportional, the true and apparent values are reversed between Eqs. 3.10 and 3.12.

You can also find an expression for the fractional change in frequency as follows:

$$\frac{f_{true}}{f_{app}} = 1 + \frac{v_{rec}}{c},$$

$$f_{true} = f_{app}\left(1 + \frac{v_{rec}}{c}\right) = f_{app} + f_{app}\left(\frac{v_{rec}}{c}\right),$$

$$f_{true} - f_{app} = f_{app}\left(\frac{v_{rec}}{c}\right),$$

$$\frac{f_{true} - f_{app}}{f_{app}} = \frac{v_{rec}}{c}.$$

If the change in frequency Δf is designated as $(f_{app} - f_{true})$, then $(f_{true} - f_{app})$ is $-\Delta f$ and this becomes

$$\frac{-\Delta f}{f_{app}} = \frac{v_{rec}}{c},$$

or

$$\frac{\Delta f}{f_{app}} = -\frac{v_{rec}}{c}. \tag{3.13}$$

Make sure you note the two differences between Eqs. 3.11 and 3.13. The first is that the frequency version has the apparent frequency (f_{app}) in the denominator, while the wavelength version has the true wavelength (λ_{true}) in the denominator. The second difference is that the frequency version has a minus sign in front of v_{rec}, since the apparent frequency (f_{app}) is smaller than the true frequency (f_{true}) if the velocity of recession is positive (that is, if the objects are moving apart).

In the next section, you can see how the Doppler effect is used in one of the most active research areas in contemporary astronomy – the search for extrasolar planets. But before moving on to that, you may want to make sure you're solid on the mathematics of the Doppler shift by working through the following exercises.

Exercise 3.10. What is the frequency shift (Δf) of an 18-GHz signal sent from a source receding from the observer at a speed of 150 km/s?

Exercise 3.11. Find λ_{app} and $\Delta\lambda$ for a star moving toward the Earth at a speed of 300 km/s while emitting light with wavelength of 530 nm.

3.4 Radial-velocity plots

The Doppler effect is the basis for one of the most effective tools used by astronomers searching for planets orbiting around stars other than the Sun. Detecting such "extrasolar" planets directly by visual observation is exceptionally difficult – imagine trying to see the light reflecting from a mosquito flying near a searchlight. But ever since Newton modified Kepler's Laws, astronomers have been aware of the possibility of detecting extrasolar planets indirectly, by measuring the effect of a planet's gravitational pull on its parent star.

That effect comes about from the second modification Newton made to Kepler's Laws. As explained in Section 2.3, Newton's first modification to Kepler's Laws was to include a mass term in the equation relating orbital period to semi-major axis. But Newton's second modification was equally important: applying his laws of mechanics to orbital motion, Newton determined that planets do not orbit the center of a stationary Sun.

Instead, both the planet and the Sun orbit a point called the "center of mass" of the Sun–planet system. You can see a conceptual depiction of this in Figure 3.10, in which the small "×" represents the center of mass between an orbiting planet and its parent star. Notice that in addition to the planet's orbit, another (smaller) orbit is shown, and that orbit is labeled "star's orbit" in the figure. The center of mass is at the focus of both the planet's orbit and the star's orbit, and the star as well as the planet orbit that point in accordance with

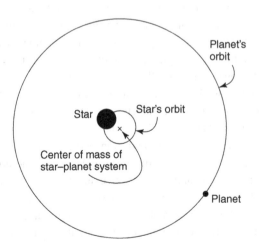

Figure 3.10 A planet and its parent star orbiting their common center of mass (not to scale).

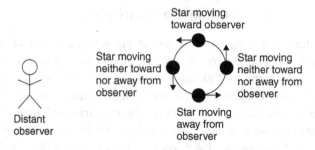

Figure 3.11 The orbital motion of a star relative to a distant observer.

Kepler's Laws as modified by Newton. The center of mass of the star–planet system is always on a line between the center of the star and the center of the planet, which means the star and planet have the same orbital period.

In reality, stars are so much more massive than planets that in many cases the center of mass between the star and the planet is located at a point *inside* the star. So rather than orbiting around an external point, the star wobbles around a point somewhere between[9] its center and its outer layer.

Once you comprehend the idea that stars with planets orbit (or at least wobble) around the center of mass, the next step to understanding radial-velocity plots is to imagine an observer watching such a star, as shown in Figure 3.11.

In this figure, the star's orbit and the observer both lie in the same plane (the plane of the page). As you can see, there is one location at which the star is moving directly toward the observer and another location at which the star is moving directly away from the observer (that is, along the line of sight). Additionally, there are two locations at which the star is moving neither toward nor away from the observer, but purely "sideways" (astronomers call this direction "transverse to the line of sight"). At all other locations, the star is moving partially toward or away and partially sideways.

Why is this important? Because astronomers use the Doppler effect to detect the orbital motion of stars with planets, and Doppler shift is produced by "radial" motion but not "transverse" motion.

The difference between radial and transverse motion is explained in Figure 3.12. Notice that the radial direction is defined as the direction along the observer's line of sight, and the transverse direction is defined as the direction perpendicular to the observer's line of sight. If an object is moving in a direction that is neither entirely radial nor entirely transverse, its total velocity can be broken down into two components: the radial velocity (along the line of

[9] The center of mass between two objects is always closer to the center of the more-massive object – you can see how to calculate its location on the book's website.

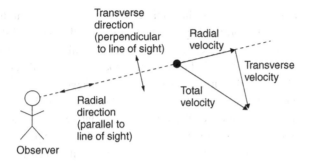

Figure 3.12 The radial and transverse velocity components.

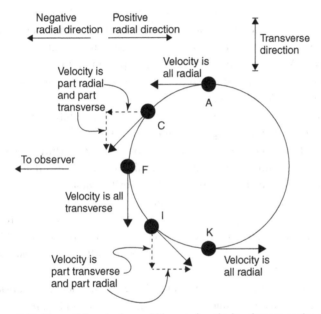

Figure 3.13 The radial part of an orbiting star's velocity changes as the star's total velocity changes direction (the solid arrows represent the star's total velocity, the horizontal dashed arrows represent the star's radial velocity, and the vertical dashed arrows represent the star's transverse velocity).

sight) and the transverse velocity (perpendicular to the line of sight). By convention, the radial velocity is considered positive if the object is moving away from the observer and negative if the object is moving toward the observer.

Radial and transverse velocity are important to understanding the observations of orbiting stars, as illustrated in Figure 3.13.

Notice that as the star orbits, the total velocity points in different directions (always tangent to the orbit). That means that even if the star's orbital speed remains the same (as would happen in a perfectly circular orbit, according to Kepler's Second Law), the radial part of the velocity is constantly changing.

Figure 3.13 depicts the circular orbit of a star (exaggerated in size, so it is orbiting a point well outside itself) as observed by a distant observer, who is far off the left side of the page. The observer is so far away that the direction toward him or her is directly to the left from anywhere on the page, regardless of where the star is in its orbit. At the top of the orbit shown in this figure (position A), the star is moving directly toward the observer, so the velocity is entirely radial (and negative). But as the star moves along its orbit to position C, the radial part of the velocity becomes smaller (that is, a smaller negative number), even though the total velocity has remained the same. And when the star reaches position F, the star's motion is neither toward nor away from the observer, so the velocity is entirely transverse (the radial part is zero) to the observer. As the star continues orbiting to position I, the radial part of the velocity grows again (this time in the positive direction), reaching its maximum when the star is at position K.

The relevance of the changing radial velocity of an orbiting star is that the Doppler shift of the star's light will go from a relatively strong blueshift at position A, to a smaller blueshift at position C, to zero shift at position F, and then small redshift at position I and relatively strong redshift at position K (remember, the Doppler shift is produced only by the *radial* part of the motion). As the star works its way around the other side of the orbit, from position K back to position A in Figure 3.13, the Doppler shift will go from relatively strong redshift to smaller redshift, then zero shift, small blueshift, and back to relatively strong blueshift. This changing Doppler shift is the hallmark of a star being orbited by a planet.

To understand the exact nature of that changing Doppler shift, consider the star's radial velocity at the intermediate positions shown in Figure 3.14. In this figure, the lengths of the dashed horizontal arrows indicate the magnitude of the radial velocity only. In order to make it easier to compare the radial velocity at the intermediate positions between A and K, the right side of this figure shows the radial velocity at each position starting from the same left–right position.

Even more instructive are the plots in Figure 3.15, in which the spacing between the positions A through K have been adjusted to represent equal intervals of time. Notice that in the left portion of this figure, the vertical spacing between position A and position B has been increased, as has the spacing between position J and position K, to represent the longer time it takes the star

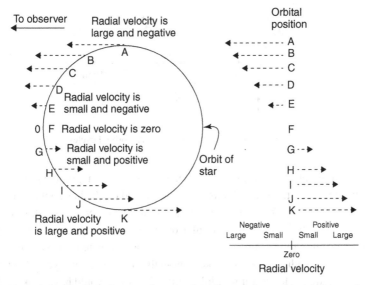

Figure 3.14 The dashed arrows show the changing radial part of an orbiting star's velocity. Not shown is the star's total velocity, which is constant and always tangent to the star's orbit.

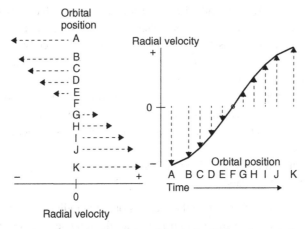

Figure 3.15 Plotting an orbiting star's radial velocity (RV) on the vertical axis and time on the horizontal axis gives a standard RV plot.

to move between these positions (because positions A and B are separated by 30°, as are positions J and K, whereas all other positions are separated by 15°.

Turning this graph on its side (by rotating 90° anti-clockwise) gives the plot shown on the right side of Figure 3.15, which has the form of a standard

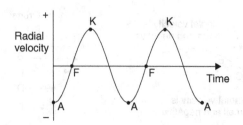

Figure 3.16 RV curve over two complete orbital cycles.

radial-velocity plot (sometimes abbreviated RV plot): a graph with radial velocity plotted on the vertical axis and time plotted on the horizontal axis.

The dark line connecting the ends of the radial-velocity arrows shows that the radial velocity of a star in a circular orbit varies sinusoidally (that is, like a sine or cosine graph). Since only one-half of a complete orbit is plotted in this figure, the radial velocity makes one-half of a complete cycle. Had we plotted the radial velocity for multiple orbits, the graph would have looked like Figure 3.16, with multiple cycles.

Notice that the starting point of a radial-velocity plot can be positive, negative, or zero, depending on the position of the star in its orbit when the first observation is made. You should also be aware that in radial-velocity plots made using data from real stars, the peaks and troughs may not be perfectly symmetrical, which means that this star's orbit is not perfectly circular but somewhat elliptical. The higher the eccentricity of the star's orbit, the more asymmetric the RV plot. Additionally, real RV plots are not continuous curves, but rather a series of individual data points to which a curve is fitted. Each data point in a radial-velocity plot represents one observation of the star during which the star's radial velocity was measuring by the Doppler shift in the star's spectrum.

When interpreting radial-velocity plots, it's very important to keep in mind the following two caveats:

(1) Since the velocity shown on the graph is only the radial component of the velocity (along your line of sight), it is not the total velocity of the object through space.
(2) Zero radial velocity on the graph does not mean the object is not moving; it means that the object is at a point in its orbit in which the velocity is entirely transverse (perpendicular to the line of sight).

Example: In the example RV graph in Figure 3.17, how many separate times was the star measured to be moving away from the observer at greater than 20 m/s?

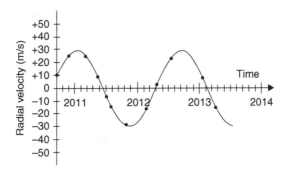

Figure 3.17 Example RV curve with large tick marks at the start of each year.

On this graph there are 12 individual data points representing 12 measurements of the star's velocity over the course of about 3 years. The smooth curve connecting the points is the "curve of best fit" (meaning that it's the sinusoidal curve that gets closest to all of the data points); this curve shows how the radial velocity changed between the times of the measurements. The star's radial velocity varies from about −30 m/s to about +30 m/s, which you can tell by looking at the y-values of the data points and the best-fit curve.

Since positive radial velocity corresponds to motion away from the observer, to answer the question asked in this example you're looking for data points at which the radial velocity was greater than +20 m/s. As you can see by the y-values of the data points on the graph there are three separate times that the star's radial velocity was measured to be above +20 m/s: one in late 2010, one in early 2011, and one in mid 2012.

The period of a planet's orbit can also be determined from an RV plot. To see how that works, consider the repeating pattern of the negative–positive velocity oscillation in an RV plot. The time from one peak in the curve to the next – or, equivalently, from one trough to the next – is the period of the orbit. That is, it is the time elapsed during exactly one full orbit of the star. For the star shown in the example RV plot in Figure 3.17, its period is about 17 tenths of a year (17 small tick marks on the x-axis), or about 20 months.

In addition to revealing the presence of a planet orbiting a wobbling star, radial-velocity measurements can also be combined with other information to determine the semi-major axis of the planet's orbit and even the mass of the planet. To find the semi-major axis, it's necessary to know the mass of the planet's parent star, which may be estimated from its location on the Hertzsprung–Russell diagram as described in Section 5.4. With the star's mass (M) in hand and the planet's orbital period (P) determined from the radial-velocity plot, the semi-major axis (a) of the planet's orbit may be found

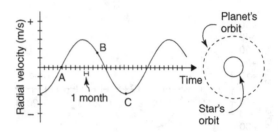

Figure 3.18 RV curve for Exercises 3.12 through 3.14.

using Newton's modified version of Kepler's Third Law ($P^2 = a^3/M$). For circular orbits, knowing the semi-major axis (equivalent to the radius for a circular orbit) and period, the planet's orbital speed may be determined by dividing the circumference of the orbit by the orbital period. And knowing the star's orbital speed, the planet's orbital speed, and the mass of the star, the mass of the planet may be determined using the law of the conservation of momentum (if you're interested, you can see how that works on the book's website).

Exercise 3.12. For points A, B, and C on the radial-velocity curve shown in Figure 3.18, identify where the star is in its orbit (the planet and the star both orbit clockwise).

Exercise 3.13. What is the orbital period of the planet orbiting the star with the radial-velocity curve shown in Figure 3.18?

Exercise 3.14. If the star in the previous exercise has a mass of 2.5 solar masses, what is the semi-major axis of the planet's orbit?

You may be wondering whether radial-velocity plots are useful when the observer is not in the plane of the star's orbit. To understand the effect of this, imagine an extreme case in which the observer's line of sight is perpendicular to the plane of the orbit as in the top portion of Figure 3.19 (in this figure, the plane of each orbit is perpendicular to the plane of the page).

For any such observer, looking at the orbit face-on, the star's motion would be everywhere perpendicular to the observer's line of sight. That observer would detect no Doppler shift from the star's motion and would correctly conclude that the star had no *radial* velocity. For such orbits, radial-velocity plots register a constant radial velocity of zero, and the radial-velocity technique is not useful in detecting wobbling stars – and thus planets – in these cases.

So, while an observer looking face-on to the star's orbit will measure zero Doppler shift, an observer in the same plane as the orbit (that is, looking at

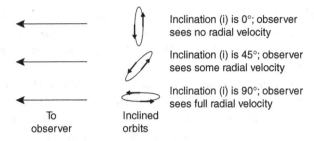

Figure 3.19 The effect of orbital inclination on radial velocity.

the orbit edge-on) will detect the full strength of the Doppler shift from the star's motion. And what of observers at intermediate angles between the plane of the orbit and the perpendicular direction? They will see a "diluted" Doppler shift, and their radial-velocity plots will have reduced amplitude (lower peaks and shallower troughs) than those of the in-plane observer. But with the radial velocity method the observer cannot know a priori at what angle they are observing the orbit because they cannot see the orbit directly; they are infer-ring the orbit from the Doppler shift of the star's light. So they typically do not know if the radial-velocity curve from their observations shows the full amplitude or a reduced amplitude. As a result, there is some ambiguity in the inferred mass of the planet that is tugging on the star and causing it to wobble.

The effect of orbital inclination is the reason you may hear planet masses derived from the radial-velocity method referred to as "minimum masses." These are often labeled as "$M \sin i$", where M is the unknown planet mass and i is the unknown angle of inclination of the orbit, which may have a value any-where between $0°$ and $90°$. This means that the trigonometric function "$\sin i$" yields a multiplicative factor between 0 and 1 (but never greater than 1). If $i = 0°$, the orbit is viewed face-on so no Doppler shift would be detected, and $M \sin 0° = 0$ corresponds to "no planet found." If $i = 90°$, the orbit is viewed edge-on and $M \sin 90° = M$, so the full planet mass is inferred.

Another way to understand this is to realize that the mass of the planet inferred by this technique ($M_{inferred}$) is the actual mass of the planet (M_{actual}) times $\sin i$:

$$M_{inferred} = M_{actual} \sin i,$$

which means that

$$M_{actual} = \frac{M_{inferred}}{\sin i}. \tag{3.14}$$

So only in the case of $i = 90°$ does M_{actual} equal $M_{inferred}$, since $\sin 90° = 1$. In all other cases the factor of $\sin i$ in the denominator is less than 1, which makes M_{actual} bigger than $M_{inferred}$. Hence $M_{inferred}$ is the minimum possible mass of the planet.

In rare cases, a planet is observed to "transit" its parent star. That means that the orbital plane is aligned so precisely with our viewing angle that the planet passes directly in front of its star once per orbit, temporarily blocking a tiny bit of the star's light, so the viewing angle is known to be very close to edge-on. In such cases the radial-velocity plot is known to show nearly the full Doppler shift, and the inferred planet mass is known to be nearly the full planet mass.

Exercise 3.15. Using the radial-velocity technique, astronomers infer the minimum mass of a planet is $0.8M_{Jupiter}$. If the inclination angle of the planet's orbit is 45°, what is the actual mass of the planet?

3.5 Chapter problems

3.1 You are listening to a radio station that broadcasts at 99.5 on your FM dial – that is, using radio waves with frequency 99.5 MHz.

(a) What is the wavelength of these waves?

(b) What is the energy of one photon of this "radio light"?

3.2 FM radio frequencies in many countries range from 88 to 108 MHz. AM radio frequencies range from 530 to 1700 kHz.

(a) Without doing any calculations, which wavelengths are longer, FM radio waves or AM radio waves?

(b) Without doing any calculations, which wavelengths travel at a faster speed, FM radio waves or AM radio waves?

3.3 How much more power does a person give off when they are running a high fever compared to when their temperature is normal? Assume a person is an ideal thermal (blackbody) radiator. Assume a normal body temperature of 310.15 K (98.6 °F; 37 °C), and a feverish temperature of 313.7 K (105 °F; 40.55 °C). Express your answer as a percent increase.

3.4 You are observing the spectrum of a star. You recognize a spectral line of hydrogen that normally occurs at a wavelength of 656 nm, but from this star it occurs at 650 nm. How fast and in what direction is the star moving relative to the Earth?

3.5 If a certain blackbody appears brightest at a frequency of 7×10^{13} Hz, what is the temperature of that blackbody?

3.6 How many times hotter or cooler is a blackbody whose radiation peaks at $\lambda = 630$ nm than a blackbody whose radiation peaks at $\lambda = 490$ nm?

3.7 In the Summer Triangle, the star Vega has $L \approx 40\ L_{Sun}$ and radius 2.3 R_{Sun}, while the star Deneb has $L \approx 200,000\ L_{Sun}$ and radius $\approx 200\ R_{Sun}$. Which of these stars is hotter, and by what factor?

3.8 Interplanetary spacecraft travel at sufficiently high speeds that the Doppler shift of their radio signals must be taken into account when designing their communication systems. For a spacecraft traveling away from Earth at a speed of 98,000 miles per hour, to what frequency should a receiver on Earth be tuned in order to receive a signal transmitted from the spacecraft at a frequency of 7.5 GHz?

3.9 If the planet orbiting the star with edge-on RV curve of Figure 3.17 has a circular orbital speed of 500 km/s, what is the mass of the star?

3.10 If the star with RV curve shown in Figure 3.17 has a transiting planet, in which months and years would transits be observed?

4

Parallax, angular size, and angular resolution

One of the most important contributions that the science of astronomy has made to human progress is an understanding of the distance and size of celestial objects. After millennia of using our eyes and about four centuries of using telescopes, we now have a very good idea of where we are in the Universe and how our planet fits in among the other bodies in our Solar System, the Milky Way galaxy, and the Universe. Several of the techniques astronomers use to estimate distance and size are based on *angles*, and the purpose of this chapter is to make sure you understand the mathematical foundation of these techniques. Specifically, the concepts of parallax and angular size are discussed in the first two sections of this chapter, and the third section describes the angular resolution of astronomical instruments.

4.1 Parallax

Parallax is a perspective phenomenon that makes a nearby object appear to shift position with respect to more distant objects when the observation point is changed. This section begins with an explanation of the parallax concept and proportionality relationships and concludes with examples of parallax calculations relevant to astronomy.

4.1.1 Parallax concept

You can easily demonstrate the effect of parallax by holding your index finger upright at arm's length and then observing that finger and the background behind it with your left eye open and your right eye closed. Now close your left eye and open your right eye, and notice what happens – your finger appears to

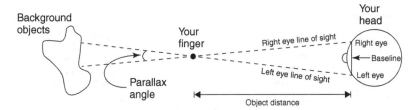

Figure 4.1 Parallax demonstration (top view).

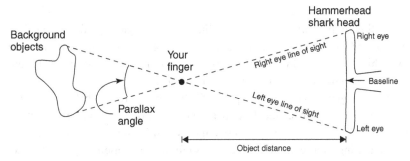

Figure 4.2 Longer baseline increases parallax angle.

shift its position with respect to the background, as illustrated in Figure 4.1. The amount of that shift, measured as an angle, is called the parallax angle.[1]

With a little thought, you should be able to convince yourself that the amount of parallax (that is, the size of the angle) in this little demonstration depends on two things: the length of your arm and the distance between your eyes. If that's not clear to you, consider what you'd see if your eyes were spaced much farther apart like those of a hammerhead shark. With very widely spaced eyes, the lines of sight from your left eye to your finger and from your right eye to your finger would be very different. And those very different lines of sight would make the parallax angle *bigger* for an object at a given distance, as shown in Figure 4.2.

By comparing the parallax angles in Figures 4.1 and 4.2, you can see that greater separation between observation points increases the parallax angle for an object at any given distance. The separation between observation points is called the "baseline" of the parallax measurement, and for small angles the parallax angle is directly proportional to the baseline. Why is this strictly true only for small angles? Because it's really the *tangent* of the parallax angle

---

[1] Some astronomy texts define the parallax angle as *half* of the angle by which the object shifts, as described in Section 5.1.

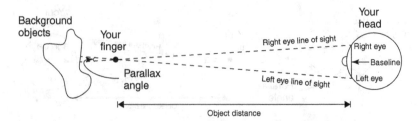

Figure 4.3 Greater distance to object decreases parallax angle.

that's directly proportional to the baseline. But for angles less than about 10°, the tangent of an angle and the value of the angle (in radians) are the same to within about 1%. In astronomy, most of the angles you'll encounter are fractions of a degree, so this approximation is useful.

Now imagine what would happen if your arm were much longer, which would make your finger much farther from your eyes when you fully extend your arm. In this case, the parallax angle would be *smaller*, as you can see by comparing Figure 4.3 to Figure 4.1. So for a given baseline, greater object distance results in smaller parallax angle. For small angles, the parallax angle is inversely proportional to the object distance.

4.1.2 Calculating parallax angle

Combining the parallax angle's direct proportionality to baseline and inverse proportionality to object distance allows you to write the following proportionality relationship:

$$\text{parallax angle} \propto \left(\frac{\text{baseline}}{\text{object distance}} \right) \tag{4.1}$$

or

$$\text{parallax angle} = (\text{const}) \times \left(\frac{\text{baseline}}{\text{object distance}} \right) \tag{4.2}$$

in which the constant of proportionality will be 1.0 as long as the units of the baseline are the same as the units of the object distance and the units of the parallax angle are radians.

Since Eq. 4.1 is a proportionality relationship, it's well suited for use with the ratio method, as shown in the following example.

Example: If switching from your left eye to your right eye produces a parallax angle of 0.04 radians for your finger at arm's length, what parallax angle

*would a hammerhead shark observe for your finger at the same distance if the
shark's eyes are three times farther apart than yours?*

Since parallax angle is directly proportional to baseline, and the shark's baseline is three times greater than yours, you can solve this problem in your head.
If the parallax angle you observe is 0.04 radians and the shark's longer baseline
increases the parallax angle by a factor of three, the parallax angle observed
by the shark will be 0.04 × 3 radians, which is 0.12 radians. Notice that you
can determine this angle without knowing values for your baseline, the shark's
baseline, or the distance of your finger, as long as you know the *ratio* of the
baselines and providing that the object distance remains the same.

   If you prefer to calculate the parallax angle in degrees, you can build the
conversion factor from radians to degrees right into the parallax equation:

$$\text{parallax angle (deg)} = 57.3° \left(\frac{\text{baseline}}{\text{object distance}} \right) \qquad (4.3)$$

in which the factor of 57.3 (which is $180/\pi$) does the conversion from radians
to degrees (because $180° = \pi$ radians). But it's important to remember that
this equation works only when the units of baseline are the same as the units
of the object distance.

   The following example illustrates the use of Eq. 4.3 to find the distance to
an object by measuring the object's parallax angle over a known baseline.

*Example: Using a baseline of 20 meters, a surveyor measures a parallax angle
of 2.4° for a distant tree. How far away is the tree?*

Since you're trying to find an object's distance using parallax, and you're given
the baseline and parallax angle, Eq. 4.3 is a good place to start. You could
just begin plugging in values but, as always, a better approach is to first rearrange the equation to isolate the quantity you're seeking (object distance) on
the left side:

$$\text{parallax angle (deg)} = 57.3 \left(\frac{\text{baseline}}{\text{object distance}} \right),$$

$$\text{object distance} = 57.3 \left(\frac{\text{baseline}}{\text{parallax angle (deg)}} \right).$$

Plugging in your values for baseline and parallax angle gives

$$\text{object distance} = 57.3° \left(\frac{20 \text{ m}}{2.4°} \right)$$

$$= 477.5 \text{ m}$$

in which the units of the object's distance are the same as the units of the baseline (meters in this case).

This example demonstrates why parallax is such a powerful tool for astronomers: using two measurable parameters (parallax angle and baseline length), the distance to remote objects may be determined. Of course, the immense distances to other planets and stars means that the parallax angles for such objects are extremely small, since distance appears in the denominator of Eqs. 4.1 and 4.3. Thus the only way to obtain measurable parallax angles for astronomical objects is to use very long baselines. One way to achieve a long baseline (approaching the diameter of the Earth) is to allow Earth's rotation on its axis to move the observation location over the course of a day. Even longer baselines (up to 2 astronomical units (AU), the diameter of the Earth's orbit) can be achieved using the Earth's revolution around the Sun, and Section 5.1 shows how this can be used to find the distances to nearby stars.

Exercise 4.1. If the minimum measurable parallax angle for a certain instrument is 10 arcseconds, what is the maximum distance at which that instrument can measure an object's parallax if a baseline equal to the Earth's diameter is used?

Exercise 4.2. Repeat the previous exercise for the case in which the minimum measurable parallax angle is 1 arcsecond.

Exercise 4.3. Repeat the two previous exercises using the diameter of the Earth's orbit as your baseline.

4.2 Angular size

Every child knows that an object "gets smaller" as it gets farther away. Yet the physical size of the object is not changing, so what makes it appear smaller? The answer is that the "angular size" of the object decreases with distance, and angular size is the subject of this section.

4.2.1 Angular-size concept

To understand why angular size decreases with distance you have to realize that the angular size (also called the "angular diameter") of an object for a given observer is the angle between the line of sight from the observer to one edge of the object and the line of sight to the opposite edge of the object. This is shown in Figure 4.4.

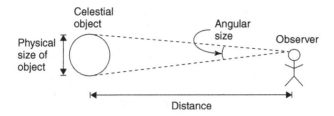

Figure 4.4 Angular size of a celestial object.

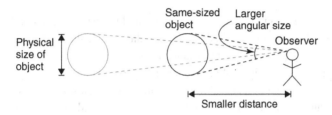

Figure 4.5 Angular-size dependence on distance.

Some astronomy students have trouble differentiating angular size from parallax angle, but you can understand the difference by comparing Figure 4.4 to
Figure 4.1. Both angular size and parallax involve the difference between two
lines of sight, but the way those lines of sight are formed is completely different. The key difference is that parallax occurs when the same object is observed
from two different locations (the ends of the baseline, as shown in Figure 4.1),
while angular-size measurement is made from the same location, but two different points on the object are observed (as shown in Figure 4.4). Put simply,
the observation point changes for parallax measurements, but the observation
point remains the same for angular-size measurements.

A tip that some students find helpful for differentiating angular size from
parallax is to remember which way the triangle points. In parallax measurements, the baseline (the base of the triangle) is between the two observing
locations, and the triangle points away from the observer. In angular-size measurements, the base of the triangle is the physical size of the object, and the
triangle points toward the observer.

Unlike physical size, which is completely determined by the dimensions of
the object, angular size depends on both the physical size of the object and the
distance to the observer. To see that, consider what would happen if the object
in Figure 4.4 were closer to the observer, as shown in Figure 4.5.

In this case, the smaller distance to the object would increase the angle
formed at the observation point by the two lines of sight to opposite edges of

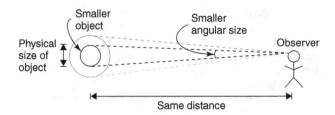

Figure 4.6 Angular-size dependence on physical size.

the object. Thus angular size is inversely related to distance; for small angles, cutting the distance in half doubles the angular size of the object (so angular size is inversely proportional to distance for small angles).

As you can see in Figure 4.6, angular size also depends on the physical size of the object – larger objects subtend a greater angle than smaller objects at the same distance. And as you may have guessed, for small angles, angular size is directly proportional to physical size. Combining the effects of distance and physical size, the angular size of an object may be written as

$$\text{angular size} \propto \left(\frac{\text{physical size}}{\text{distance}} \right) \qquad (4.4)$$

or

$$\text{angular size} = (\text{const}) \times \left(\frac{\text{physical size}}{\text{distance}} \right). \qquad (4.5)$$

4.2.2 Calculating angular size

The constant of proportionality in Eq. 4.5 is 1.0 as long as the units of the angular size are radians and the units of the physical size are the same as the units of the distance. But even without knowing the units, you can use the proportional relationship of Eq. 4.4 or 4.5 to solve angular-size problems using the ratio method. Here's an example.

Example: The Sun's diameter is approximately 1,390,000 km, which is about 400 times larger than the Moon's diameter. But the Sun's distance from Earth is also about 400 times larger than the Moon's distance, so how does the angular size of the Sun compare to the angular size of the Moon, as seen from Earth?

As in the shark–parallax example, you can use the ratio method to solve this problem in your head. Since physical size is in the numerator of Eq. 4.4 and distance is in the denominator, the Sun's greater size will be offset by its greater distance – if both numerator and denominator are 400 times larger for the Sun compared to the Moon, then the angular size of the Sun and the Moon must be approximately equal. That's why the Moon can just barely cover the Sun during a total solar eclipse.[2]

Here's a version of the angular-size equation that reminds you of the units:

$$\text{angular size (rad)} = \frac{\text{physical size (same units as distance)}}{\text{distance (same units as physical size)}}. \tag{4.6}$$

Example: What is the diameter of a planet whose angular size is 47″ as seen from the Earth when the distance to the planet is 4.2 AU?

In this problem, you're given the angular size of an object and the distance from the observer to the object, and you're asked to find the physical size of the object. These are the variables in Eq. 4.6, although you'll have to do some unit conversion before you can use that equation. And, as always, it's a good idea to first rearrange the equation to move the quantity you're after onto the left side:

$$\text{angular size} = \frac{\text{physical size}}{\text{distance}}$$

$$\text{physical size} = (\text{angular size})(\text{distance}). \tag{4.7}$$

Before plugging in the values for angular size and distance, you'll have to convert the units of angular size from arcseconds to radians, using the fact that $1° \leftrightarrow 3{,}600″$:

$$\text{angular size} = 47″ \times \left[\frac{1 \text{ degree}}{3{,}600″} \right] \times \left[\frac{\pi \text{ radians}}{180 \text{ degrees}} \right]$$

$$= 2.28 \times 10^{-4} \text{ rad.}$$

And since the units of AU work well for interplanetary distances but are generally not convenient for expressing the diameter of a planet, converting the distance of 4.2 AU to kilometers is also a good idea:

$$\text{distance} = 4.2 \text{ AU} \times \left[\frac{1.5 \times 10^8 \text{ km}}{1 \text{ AU}} \right] = 6.3 \times 10^8 \text{ km.}$$

---

[2] Since the Moon's orbit around the Earth and the Earth's orbit around the Sun are elliptical rather than circular, the Earth–Moon and Earth–Sun distances change slightly over time. This gives rise to "annular" solar eclipses in which the Moon's angular size is smaller than the Sun's so the Sun is not entirely covered.

With the parameters in the correct units, you can now plug the values into Eq. 4.7:

$$\text{physical size} = (\text{angular size})(\text{distance}) = (2.28 \times 10^{-4} \text{ rad})(6.3 \times 10^8 \text{ km})$$
$$= 143{,}640 \text{ km},$$

which is the diameter of the planet Jupiter.

If you want to find the angular size of an object in degrees, you can use Eq. 4.6 to find the angular size in radians and then convert your answer to degrees, or you can include the conversion factor in the angular-size equation like this:

$$\text{angular size (deg)} = \left[\frac{180 \text{ deg}}{\pi}\right] \times \text{angular size (rad)}$$
$$= \left[\frac{180 \text{ deg}}{\pi}\right] \times \frac{\text{physical size}}{\text{distance}}$$

or

$$\text{angular size (deg)} = 57.3° \left[\frac{\text{physical size}}{\text{distance}}\right], \tag{4.8}$$

in which the units of the object's physical size must be the same as the units of the distance.

The next section of this chapter will help you understand why the very small angular size of the stars makes it virtually impossible to resolve their surfaces even with the largest telescopes currently available. But first, here's a chance to exercise your understanding of angular size.

Exercise 4.4. What is the angular size of the star Betelgeuse as seen from Earth? The diameter of Betelgeuse is estimated to be 1000 Sun diameters, and the distance to Betelgeuse is about 650 light years.

4.3 Angular resolution

As described in the previous section, the immense distances of astronomical objects means that those objects have very small angular sizes, and it's hard to see things with very small angular size. But in 1608, spectacle-maker Hans Lippershey and others in the Netherlands began using lenses to make distant objects appear larger. Galileo heard about this discovery the following year, and he immediately set about making his own (greatly improved) version of what would soon be called the "telescope." He then turned the instrument to the night sky, and we've been using telescopes to observe astronomical objects with small angular size ever since.

Most comprehensive astronomy texts have a section dealing with telescopes and other astronomical instruments, and there's usually not much mathematics in those sections. But when you read about the benefits of large telescopes, you're likely to encounter terms such as "sensitivity" and "angular resolution." The concept of sensitivity is straightforward: bigger telescopes (that is, telescopes with a wider aperture – the opening through which light enters) gather more light, and this makes faint objects appear brighter. But we think angular resolution warrants additional explanation, and that's the subject of the remainder of this section.

4.3.1 Angular-resolution concept

What exactly is angular resolution? It's the minimum angle over which two points may be seen as separate rather than blurred together. Those two points may be two separate light sources, such as two stars, or they may be details on a single object, such as the edge of the Red Spot on Jupiter. So with better (smaller) angular resolution, you're able to see objects more clearly. If you wear eyeglasses or contact lenses, you can easily demonstrate the benefits of better angular resolution by comparing your view of the world with and without your lenses. Without optical aid, your eyes present you with a view of the world that is fuzzy and in which details are unresolved.

Resolution is related to the wave nature of light, and to understand resolution, you need to understand how waves interact. When two or more waves are present at the same location, the interaction between those waves is called "interference." And although interference has a negative connotation in everyday use, in science interference may be constructive, destructive, or something in-between. A few examples of interference between two waves are illustrated in Figure 4.7.

Notice that when the two waves are in step (also called "in phase"), they add constructively to produce a larger wave. But if those same two waves are out of step, they add destructively to produce a smaller wave (which may be no wave at all, if the two waves are perfectly out of phase and equal in size). And, if the waves are just slightly out of step, the resultant wave is not as big as the perfectly in-step case, but it is still bigger than either of the constituent waves.

To see why wave interference is relevant to angular resolution, you have to consider the waves gathered and brought together by a lens or mirror. In Figure 4.8, you can see a slice through a lens and the effect of the lens on incoming light waves.

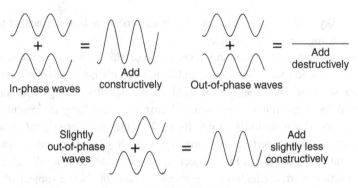

Figure 4.7 Wave interference.

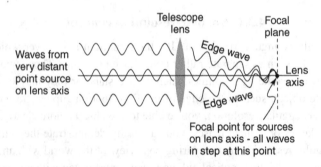

Figure 4.8 Waves passing through a lens.

Notice that in this figure, the incoming waves on the left are all parallel to one another; this is due to the very great distance to source. For a closer source, the waves would be diverging (getting farther apart as they travel), but even the closest astronomical objects are so far away that their waves are essentially parallel by the time they get to Earth.

As you can see in this figure, after passing through the lens the light waves converge toward a point (the angles are exaggerated for clarity). Since the light source in this figure is on the axis of the lens (which is the line passing through the center of the lens and perpendicular to the lens), the "focal point" (the point to which the waves converge) is also on the lens axis. Waves coming from other directions will focus to different points, and the "focal plane" is the locus of all the points to which the waves converge.

It's important to understand that at the focal point shown in Figure 4.8, waves from all points on the lens (center, edges, and in-between) are all in step. That means that these waves will add constructively at this location, producing a bright spot on the image. But even if the source of light is a point

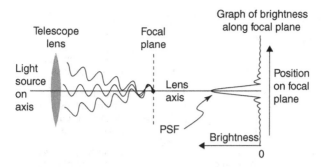

Figure 4.9 Brightness near the focal point.

(near-zero angular size), the waves add constructively not just at a single point on the focal plane, but over a small region. You can get an idea of this by looking at the little sideways graph on the right side of Figure 4.9, which is a plot of the brightness on a slice through the focal plane.

That strange-looking graph with a large central peak surrounded by lots of little bumps is called the "point-spread function" (PSF) because it shows how the light from a single source point is spread out on the image. Notice that the central peak is not infinitely narrow; it has finite width even if the light waves come from a single point. You should also notice that there are a series of "nulls" (points of zero brightness) between the minor peaks on both sides of the central peak.

The width of the main peak and the location of the nulls and minor peaks depend on two things: the size of the lens and the wavelength of the light. For a given wavelength, bigger lenses produce narrower peaks and smaller lenses produce wider peaks. And for a given lens, longer wavelengths produce wider peaks and shorter wavelengths produce narrower peaks. To understand why that's true, look at Figure 4.10.

As shown in this figure, at the point of maximum brightness, waves from all points on the lens add in-phase. Moving a small distance away from the focal point causes the waves from the edge of the lens and the waves from the center of the lens to be slightly out-of-step, so they add to a smaller value of brightness. Moving farther from the focal point causes the wave to get increasingly out-of-step, so the value of the brightness gets smaller. Eventually, if you move far enough from the focal point, the edge and center waves are completely out of step (out of phase by 180°), so they cancel. The canceling waves produce zero brightness, so the PSF has a null at this location. As you continue moving away from the focal point, some of the waves get back in phase, but others are out of phase, producing a series of minor peaks. You can see a detailed

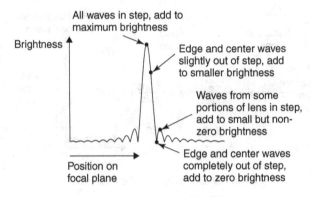

Figure 4.10 Example of a PSF.

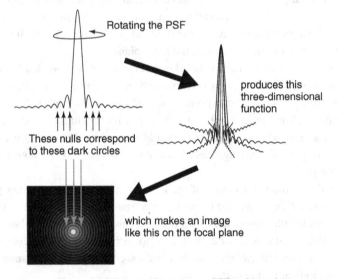

Figure 4.11 Relationship between PSF and image.

analysis of lens operation in an Optics book, but the important concept for you to understand is that even point sources don't focus to a true point – the light is spread over a small region. The bigger that region, the "fuzzier" the image looks.

To understand how to relate the graphs of the PSF to the image you see when you look through a telescope, take a look at Figure 4.11. Since the PSF graphs shown in Figures 4.9 and 4.10 represent a single slice through the focal plane produced by one slice of the lens, the entire image produced by a circular lens can be viewed as the combination of many such slices, each taken at a different

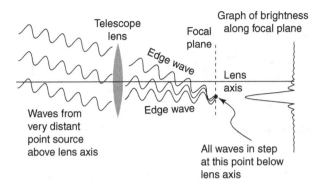

Figure 4.12 Waves from a point source above the lens axis.

angle. The rotated PSFs combine to produce the brightness function shown on the right side of Figure 4.11, which projects onto the focal plane as a bright spot surrounded by concentric rings. The spot in the middle corresponds to the bright central peak, the light rings correspond to the minor peaks of the PSF, and the dark rings correspond to the nulls of the PSF. This ring pattern is called the "Airy pattern" produced by the lens from a point source, and the bright central spot is called the "Airy disk."

So what do PSFs and Airy patterns have to do with angular resolution? To understand that, consider the waves coming from a source slightly above the lens axis, as shown in Figure 4.12.

Notice that in this case the focal point at which the waves add in phase is below the lens axis, and the peak of the PSF is shifted downward relative to the on-axis case of Figure 4.9. Likewise, for sources below the lens axis, the peak of the PSF appears above the lens axis on the focal plane.

Now consider what happens when the waves from two sources strike the lens at the same time, as shown in Figure 4.13.

In this figure, the wave directions are indicated by straight lines (called "rays") to make it easier to show waves from two directions striking the lens. As you can see in the figure, the two sources produce two PSFs, which may overlap (depending on the angular separation between the sources). Figure 4.14 shows two PSFs and the corresponding Airy disks on the focal plane. In this case, the peaks of the two PSFs are sufficiently separated to show that two separate sources exist – these two sources are said to be "resolved."

But consider a situation in which the angular separation between two sources is small enough so that their PSFs overlap significantly – not just in the minor peaks and nulls, but in the main peaks as well. If the peaks of the PSFs overlap

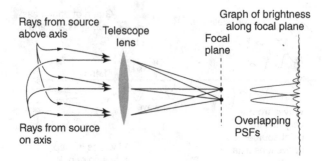

Figure 4.13 Two sources producing overlapping PSFs.

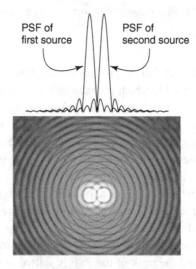

Figure 4.14 The PSFs and images of two sources.

so much that it's impossible to tell if there are two separate sources or one extended source, the sources are not resolved.

Exactly how much overlap between the peaks of the PSFs can be tolerated before the two sources become indistinguishable from a single, larger source? Several criteria exist for determining whether two sources are resolved, but most common is the Rayleigh criterion. To meet this criterion, the separation between the peaks of the two PSFs must be at least as great as the separation between the peak and the first null of a single PSF, as shown in Figure 4.15.

Meeting the Rayleigh criterion ensures that there is a small dip in brightness between the peaks, and an observer can recognize that there are two separate sources rather than a single extended one. You can see an example of an image of two sources just resolved by the Rayleigh criterion in Figure 4.16.

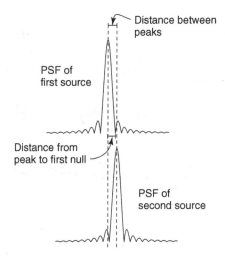

Figure 4.15 The Rayleigh criterion for resolving two sources.

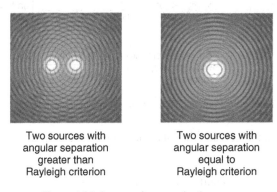

Two sources with
angular separation
greater than
Rayleigh criterion

Two sources with
angular separation
equal to
Rayleigh criterion

Figure 4.16 Images of two resolved sources.

The final concept you need to consider before you can understand the equation for angular resolution based on the Rayleigh criterion was mentioned above: the width of the PSF depends on the wavelength of the light and the size of the lens or mirror. For any given wavelength, the larger the lens, the *narrower* the PSF. So a telescope with a large aperture produces a narrower PSF than a telescope with a smaller aperture, as shown in Figure 4.17.

The reason for this is that larger lenses have greater distance from the center to the edge of the lens, and the greater that distance, the less angle it takes to cause the edge waves to get out of step with the center waves. And since bigger lenses produce narrower PSFs, the angular resolution of a big lens is better (that is, smaller) than the angular resolution of a small lens. This

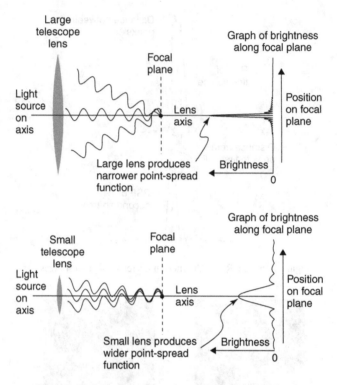

Figure 4.17 Point-spread functions of large and small lenses.

means that the angular resolution of a telescope is inversely proportional to the aperture:

$$\text{angular resolution} \propto \frac{1}{\text{aperture}}, \qquad (4.9)$$

where "aperture" is the edge-to-edge size of the lens or mirror.

The width of the PSF also depends on the wavelength of light used, because the shorter the wavelength, the smaller the angular shift it takes to make the edge waves get out of step with the center waves. This means that the angular resolution is directly proportional to wavelength (λ):

$$\text{angular resolution} \propto \lambda. \qquad (4.10)$$

4.3.2 Calculating angular resolution

Combining the proportionality relationships of Eqs. 4.9 and 4.10 gives

$$\text{angular resolution} \propto \frac{\lambda}{\text{aperture}}, \qquad (4.11)$$

or

$$\text{angular resolution} = (\text{const}) \times \left[\frac{\lambda}{\text{aperture}} \right]. \qquad (4.12)$$

If angular resolution is expressed in radians and the units of λ are the same as the unit of aperture, the constant of proportionality in Eq. 4.12 is approximately 1.22 for a circular lens. Thus

$$\text{angular resolution (rad)} = 1.22 \times \left[\frac{\lambda \text{ (same units as aperture)}}{\text{aperture (same units as } \lambda)} \right]. \qquad (4.13)$$

This is called the "diffraction-limited" resolution, and it represents the best possible resolution that a telescope could theoretically achieve – any imperfections in the lens or mirror and turbulence in the Earth's atmosphere will degrade the angular resolution.

Example: A telescope with a 10-inch lens is used to observe celestial objects. What is the angular resolution of this telescope in the middle of the visible range?

Since you're asked to find the angular resolution and you're given the aperture and wavelength (or information that allows you to determine the wavelength), Eq. 4.13 will help you solve this problem.

The wavelengths of visible light range from around 400 nm to 700 nm, so the middle of the visible range is about 550 nm. But in this problem the aperture is given in units of inches, and Eq. 4.13 requires that wavelength and aperture have the same units. You could convert nanometers to inches, but it's probably just as easy to convert both wavelength and aperture to meters:

$$\lambda = 550 \text{ nm} = 550 \times 10^{-9} \text{ m} = 5.5 \times 10^{-7} \text{ m},$$
$$\text{aperture} = 10 \text{ in} = 10 \text{ in} \times \left[\frac{1 \text{ m}}{39.37 \text{ in}} \right] = 0.254 \text{ m}.$$

Plugging these values into Eq. 4.13 gives

$$\text{angular resolution (rad)} = 1.22 \times \left[\frac{\lambda(\text{ same units as aperture})}{\text{aperture(same units as } \lambda)} \right]$$
$$= 1.22 \times \left[\frac{5.5 \times 10^{-7} \text{ m}}{0.254 \text{ m}} \right] = 2.64 \times 10^{-6} \text{ radians,}$$

which can be converted to arcseconds as follows:

$$2.64 \times 10^{-6} \text{ radians} = 2.64 \times 10^{-6} \text{ radians} \times \left[\frac{180 \text{ degrees}}{\pi \text{ radians}} \right] \times \left[\frac{3{,}600''}{1 \text{ degree}} \right]$$
$$= 0.54 \text{ arcseconds,}$$

which is about half an arcsecond or 0.00015°! But don't expect to be able to put all of this resolution to use when you're looking through the Earth's atmosphere, which typically limits the achievable resolution to 1″ or more, even under the best circumstances. So what good are telescopes larger than a few inches if the atmosphere prevents them from achieving their theoretical resolution? The answer is sensitivity – as we mentioned at the start of this section, one of the benefits of using a telescope is the ability to see fainter objects. And the larger the aperture, the more light the telescope gathers, and the brighter the image appears.

Since wavelength is usually expressed in units that are much smaller than the units of aperture, some astronomy texts use a version of Eq. 4.13 in which the units of wavelength are micrometers (μm) and the units of aperture are meters, and which gives the angular resolution in units of arcseconds:

$$\text{angular resolution (arcsec)} = 0.25 \left[\frac{\lambda(\mu m)}{\text{aperture (m)}} \right] \qquad (4.14)$$

Here are two exercises that will help you check your understanding of the equation for angular resolution.

Exercise 4.5. The undilated pupil of a typical human eye has a diameter of about 4 mm. Calculate the angular resolution of the human eye for visible light.

Exercise 4.6. Calculate the angular resolution of the human eye for visible light for a dilated pupil with a diameter of 9 mm.

4.4 Chapter problems

4.1 What is the parallax angle of the planet Venus using Earth's diameter as baseline when Venus is 0.5 AU from Earth?

4.2 How far away is a planet with a parallax angle of 4.6″ when a baseline of 1,000 km is used?

4.3 What length of baseline would produce a parallax angle of one arcsecond for the Andromeda Galaxy at a distance of 2.5 million light years from Earth?

4.4 Jupiter's moon Io orbits at a distance of about 420,000 km from Jupiter. What is the angular size of Io's orbit when Jupiter (and Io) are 4.2 AU from Earth?

4.5 What is the angular size of the moon Triton's orbit around Neptune as seen from Earth when Neptune is at closest approach to Earth? Triton's orbital period around Neptune is 5.9 Earth days, and Neptune's orbital period around the Sun is about 164 Earth years.

4.6 How far away is a galaxy with physical size (diameter) of 100,000 light years and angular size of 12 arcminutes?

4.7 What is the angular resolution in visible light of the Keck telescope, which has a mirror with diameter of 33 feet?

4.8 What is the angular resolution of the 100-meter Greenbank radio telescope when operated at a frequency of 400 MHz?

4.9 How large would a radio telescope operating at 1 GHz have to be to achieve the same resolution as the 200-inch Palomar telescope at visible light?

4.10 Imagine an extrasolar planet orbiting a 3-solar-mass star once every 75 Earth days. If the star is 10 light years from Earth, how large would a visible-light telescope have to be to resolve the planet from its parent star?

5

Stars

The sight of the starry night sky has been inspiring humans for thousands of years. For most of our history, the nature of those shimmering jewels has been a mystery, thought by many to be forever beyond our understanding. But in the last two hundred years, we've learned how to extract information as well as inspiration from starlight. By taking that light apart using spectrographs and making precise measurements of its brightness, we've come to understand a great deal not only about the nature of stars, but also about the structure and workings of the Universe at large. The mathematics behind that understanding is the subject of this chapter.

5.1 Stellar parallax

As described in Section 4.1, parallax is an apparent shift of an object's position due to the changing line of sight between the observer and the object. The amount of that shift depends on the distance to the object and on how far the observer moves (called the baseline of the measurement). Astronomers take advantage of this effect as Earth moves around in its orbit to measure distances to nearby stars, which appear to shift position against distant background stars.

5.1.1 Stellar parallax equation

As the Earth moves from one side of its orbit to the other over the course of half a year, a star's resulting parallax angular shift and its distance from Earth are related by this equation:

$$d = \frac{1}{p},$$ (5.1)

where d is the distance to the star in units of parsecs (pc), and p is the parallax angular shift in units of arcseconds ($''$). One parsec is a huge distance compared to human experience (1 pc $= 3.09 \times 10^{13}$ km $= 3.26$ light years), but it is relatively small in astronomy, which frequently deals with immense distances. Likewise, 1 arcsecond is a tiny angle compared to human experience ($1'' =$ 1/3,600th of 1 degree), but it is useful in astronomy, which frequently deals with very tiny angles.

If you want to see how this equation comes about from the general parallax Eq. 4.1, take a look at the problems at the end of this chapter and the on-line solutions. If you do that, you'll see that the angle used in Eq. 5.1 is not the full parallax angle shown in Figure 4.1, but rather half of that angle. Most astronomy texts use the term "parallax angle" or "parallax angular shift" to refer to the angle that is half of the full parallax angle shown in Figure 4.1, so we'll do the same.

In order for Eq. 5.1 to work, the quantities must always be in the specified units. In fact, the distance unit of 1 parsec is defined as the distance from Earth to a star which shows a parallax angular shift of 1 arcsecond as viewed from opposite sides of Earth's orbit. If you take care to ensure that your quantities are always in these prescribed units, then this equation is just an inverse proportionality with no constants, making it one of the simplest you will encounter in astronomy.

Example: Below are the parallax angles for four stars. Which of these stars is farthest from Earth, and which is closest?

Alcor: parallax angle $= 0.04''$ *Procyon: parallax angle $= 0.3''$*
Kappa Ceti: parallax angle $= 0.1''$ *GQ Lupi: parallax angle $= 0.008''$*

Remember that an inverse proportionality means that as one quantity gets smaller, the other gets larger, and vice versa. That is, the farther away a star is, the *larger* its distance, and therefore the *smaller* its parallax angle will be. So, in order to find the farthest star, look for the smallest parallax angle. Of these four stars, this is star GQ Lupi with parallax angle of $0.008''$. The closest star is Procyon, because it has the largest parallax angle of $0.3''$.

Eq. 5.1 can only be used as a distance-measuring tool for objects outside our Solar System, such as other stars. And it isn't possible to measure parallax for all stars – only the nearby ones in our own Galaxy, where "nearby" in this case means within a few hundred parsecs. At distances larger than that, although the parallax phenomenon still occurs, the angles are so tiny that even the best instruments do not have sufficient angular resolution to detect them, as described in Section 4.3.

5.1.2 Solving parallax problems: absolute method

Given either a distance or a parallax angle, Eq. 5.1 can be used to calculate
the other. If the quantity you are given has the appropriate units (parsecs for
distance or arcseconds for angle), then you can simply plug in the given value
and do the calculation of one divided by that value. Your answer will automatically come out in the correct units for the other value. If the number you are
given does not have the required units, then you must perform a unit conversion before plugging into Eq. 5.1 (for a refresher on unit conversions, take a
look at Section 1.1).

Example: How far away is the star Alcor in the previous example?

Alcor has a parallax angle of 0.04″, and since arcseconds are the correct units
for using the parallax equation, you can plug 0.04 directly into Eq. 5.1. Rewrit-
ing the parallax angle as a fraction (0.04 = 4/100) often makes the calculation
easier, and in this case it is simple enough that a calculator is not needed:

$$d = \frac{1}{p} = \frac{1}{0.04} = \frac{1}{\frac{4}{100}} = \frac{100}{4} = 25 = 25 \text{ pc.}$$

Since you plugged in a parallax angle in units of arcseconds, the distance you
calculated is in units of parsecs automatically. The shortcut employed to get
from $\frac{1}{4/100}$ to $\frac{100}{4}$, using the fact that 1 divided by any fraction is simply the
inverse of that same fraction, frequently comes in handy when doing parallax
problems.

*Example: Polaris (the "North star") is 434 light years away. What is its
parallax angle?*

This problem requires two steps before you can plug numbers into the parallax
equation. First, you should rearrange Eq. 5.1 to solve for the parallax angle p,
since p is the quantity you are asked to calculate:

$$d = \frac{1}{p} \quad \rightarrow \quad d \times \frac{p}{1} = \frac{1}{p} \times \frac{p}{1} = 1,$$

$$d \times p = 1 \quad \rightarrow \quad \frac{d \times p}{d} = \frac{1}{d},$$

$$p = \frac{1}{d}. \tag{5.2}$$

Second, the distance is given in units of light years instead of parsecs,
so you must perform a unit conversion. The relevant conversion factor is

1 pc $\leftrightarrow$ 3.26 ly. You can combine the unit conversion step with plugging the value of d into Eq. 5.2:

$$p = \frac{1}{d} = \frac{1}{434 \text{ ly}} \cdot \left(\frac{3.26 \text{ ly}}{1 \text{ pc}}\right) = \frac{3.26}{434 \text{ pc}} = \frac{3.26}{434} \text{ arcsec} = 0.0075''$$

Using the necessary distance units of parsecs in the denominator guaranteed that the answer for the angle would come out in arcseconds. This angle, about 7.5 *thousandths* of an arcsecond (or 7.5 *milli*arcseconds), is readily detectable by modern research telescopes.

Exercise 5.1. Find the distance in light years to the star GQ Lupi.

Exercise 5.2. The precision of the parallax-measuring instrument on board the Hipparcos satellite was approximately 0.002'' (two milliarcseconds). What is the greatest distance that could be measured using parallax with this instrument?

5.1.3 Solving parallax problems: ratio method

If you only need to *compare* two quantities rather than calculating an absolute number, then the ratio method can save you time and effort. This example shows how to use the ratio method to solve a parallax problem.

Example: The nearest star to Earth after the Sun is Proxima Centauri, at about 4 ly. Polaris's distance is 100 times larger than Proxima's distance. How do their parallax angles compare?

You could solve this problem using the absolute method, as in the previous example. Since you are given both distances, you could calculate both parallax angles individually and then compare them by dividing one by the other. However, the ratio method is faster because it bypasses the unnecessary intermediate step of calculating both angles.

If you remember that distance and parallax angle share an *inverse* proportionality relationship with one another, you might intuit the answer. Since Polaris's distance is 100 times *larger*, its parallax angle should be 100 times *smaller*. To solve this problem by writing it out mathematically, apply the ratio method as described in Section 1.2.3. First, write out the parallax equation separately for both stars, using subscripts to specify the star to which each equation applies:

$$d_{Proxima} = \frac{1}{p_{Proxima}} \quad \text{and} \quad d_{Polaris} = \frac{1}{p_{Polaris}}.$$

Now divide one equation by the other and simplify:

$$\frac{d_{Proxima} = \frac{1}{p_{Proxima}}}{d_{Polaris} = \frac{1}{p_{Polaris}}} \rightarrow \frac{d_{Proxima}}{d_{Polaris}} = \frac{1}{p_{Proxima}} \times \frac{p_{Polaris}}{1},$$

$$\frac{d_{Proxima}}{d_{Polaris}} = \frac{p_{Polaris}}{p_{Proxima}}. \tag{5.3}$$

Now translate the information given in the problem – "Polaris's distance is 100 times larger than Proxima's distance" – from words into a mathematical relationship, as described in Section 1.2.4:

$$d_{Polaris} = 100\, d_{Proxima}.$$

Substituting $100\, d_{Proxima}$ into Eq. 5.3 in place of $d_{Polaris}$ gives

$$\frac{d_{Proxima}}{100\, d_{Proxima}} = \frac{p_{Polaris}}{p_{Proxima}},$$

$$\frac{1}{100} = \frac{p_{Polaris}}{p_{Proxima}}.$$

Section 1.2.4 provides some guidance on interpreting this ratio answer, which tells you that the ratio of the parallax angles is 1 to 100, with Proxima Centauri's parallax angle being 100 times larger. In other words, Polaris's parallax angle is 100 times *smaller* (i.e. one-hundredth as large). Multiplying through by $p_{Proxima}$ makes this even more clear:

$$p_{Polaris} = \frac{1}{100}\, p_{Proxima}$$

This result agrees with the prediction that Polaris's parallax is 100 times smaller than Proxima's. Notice that the ratio method emphasizes the *comparison* between values, so you didn't need to use the individual distances, only their ratio.

Exercise 5.3. Use the ratio method to calculate how many times farther Kappa Ceti is than Proxima Centauri.

5.2 Luminosity and apparent brightness

If you think about it, the question "How bright is that object?" can have two meanings. It can mean "How much light is that object giving off?" or it can mean "How much light are you receiving from that object?"

The first of these questions deals with the intrinsic brightness of the object, which astronomers call "luminosity." The luminosity of an object is defined as the amount of power radiated by an object, with SI units of watts. When the label on a lightbulb says "60 watts," that label is telling you the luminosity of the bulb.

The second question deals with an observer's perception or measurement of the intensity of the light coming from an object. Two 60-W lightbulbs have the same luminosity, but if you hold one of them a few inches in front of your face and put the other one a kilometer away, the more-distant bulb will appear much dimmer. That's because the light spreads out from the bulbs in all directions, and a smaller fraction of those 60 watts from the distant bulb make it into the pupils of your eyes (or the aperture of your measuring instrument). The relevant quantity for an observer is called the "apparent brightness" of the light source, and it involves both the luminosity of the object and the distance between the source and the observer. The SI units of apparent brightness are watts per square meter. You can see an illustration of luminosity and apparent brightness in Figure 5.1.

The reason that the standard units of apparent brightness are watts per square meter is that apparent brightness is a measure not of power, but of *received energy flux*, which is called "power density" by engineers. This is a very useful quantity, because it tells you the amount of power (with SI units of watts) spread out over a certain area (with SI units of square meters). Received energy flux depends only on the luminosity of the source and the distance to the observer, so, at any given distance, all observers at that distance will receive the same energy flux. The actual received power for any one observer depends on the aperture of the observing system, which may differ from one observer to the next.

Think of it this way: if the energy flux at some distance from a light source is 20 microwatts per square meter and your optical system has an aperture

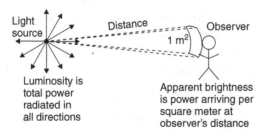

Figure 5.1 Luminosity and apparent brightness.

area of one-tenth of a square meter, you'll receive one-tenth of 20 microwatts, which is 2 microwatts. But another observer at your same distance who has an optical system with a smaller aperture area of one-twentieth of a square meter will gather in one-twentieth of 20 microwatts, which is 1 microwatt. So the received power (in watts) depends on the observer's optical system, but the energy flux (in watts per square meter) is the same for all observers at the same distance.

This means that energy flux at a given distance is a more fundamental quantity than power received by an observing system, which depends on the aperture of the system. If you encounter a situation in which you want to know the power received by a certain optical system, you can simply multiply that energy flux by the system's aperture area.

You may be wondering exactly how the received energy flux changes over distance, and the answer is a very important concept in physics. Experiments have shown that the intensity of radiation varies inversely with distance (that is, as distance increases, intensity decreases). That fits with common sense, which tells you that the farther away you move from a light source, the dimmer it appears. But the decrease in intensity is not just inversely proportional to distance from the source, it's inversely proportional to the *square* of the distance. So if you're at some distance from a light source and your friend is twice as far away, the energy flux at her location will be one-quarter (not one-half) the energy flux at your location (because $\frac{1}{2^2} = \frac{1}{4}$). And if her distance from the source is three times farther than yours, she'll measure an energy flux that is one-ninth of yours (since $\frac{1}{3^2} = \frac{1}{9}$). This is an example of the "inverse-square law": inverse because intensity decreases with distance, and square because intensity decreases as distance to the second power.

So what does this mean about the apparent brightness? Well, since apparent brightness is energy flux, and since energy flux decreases with distance according to the inverse-square law, you can write

$$\text{apparent brightness} \propto \frac{1}{dist^2}, \tag{5.4}$$

where *dist* is the distance from the light source to the observer.

Experiments also reveal that power density is directly proportional to the luminosity of the light source. So doubling the luminosity (L) of a source doubles the power density at any given range from that source. Thus

$$\text{apparent brightness} \propto L. \tag{5.5}$$

Combining the proportionality relationships of Eqs. 5.4 and 5.5 gives

$$\text{apparent brightness} \propto \frac{L}{dist^2} \tag{5.6}$$

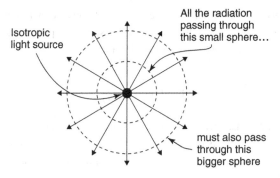

Figure 5.2 Radiation from isotropic light source.

or

$$\text{apparent brightness} = \text{const} \times \left(\frac{L}{dist^2}\right). \qquad (5.7)$$

The constant of proportionality in Eq. 5.7 can be determined by considering the case of a light source that radiates equal power in all directions; such a source is called "isotropic." The radiation from an isotropic light source is shown in Figure 5.2.

In this figure, the dashed circles represent two imaginary spheres surrounding the isotropic light source, and the arrows represent radiation emanating from the source in all directions (three-dimensionally) outward. As long as none of the light is absorbed in the space between the spheres (often a good assumption in the near-vacuum of interstellar space), all of the light that passes through the smaller sphere must also pass through the larger sphere. So the total power (the number of watts) hitting both spheres must be the same. But now consider the energy flux on each sphere. Since energy flux is defined as the power per unit area, the apparent brightness of the source for an observer at the inner (smaller) sphere is

$$(\text{apparent brightness})_{\text{inner}} = \frac{L}{SA_{inner\ sphere}}, \qquad (5.8)$$

where SA = the surface area. Since the surface area of a sphere is 4π times the radius (R) of the sphere squared, this is

$$(\text{apparent brightness})_{\text{inner}} = \frac{L}{4\pi(R_{inner})^2}. \qquad (5.9)$$

Likewise, for an observer at the outer (larger) sphere, the apparent brightness is

$$(\text{apparent brightness})_{\text{outer}} = \frac{L}{4\pi(R_{outer})^2}. \qquad (5.10)$$

But for an observer on either of these spheres, or any other imaginary sphere surrounding the light source, the radius of the sphere is just the distance from the source to the observer. So both Eq. 5.9 and Eq. 5.10 can be written more generally as

$$\text{apparent brightness} = \frac{L}{4\pi (dist)^2}. \tag{5.11}$$

Thus the constant in Eq. 5.7 is $\frac{1}{4\pi}$ as long as the source radiates isotropically, as many astronomical objects do. The standard units for the quantities in Eq. 5.11 are watts for luminosity, meters for distance, and watts per square meter for apparent brightness.

Example: The Sun's luminosity is approximately 4×10^{26} W and its distance from Earth is about 150 million kilometers. Ignoring reflection by clouds and absorption in the Earth's atmosphere, what is the apparent brightness of the Sun at the surface of the Earth?

In this problem, you're given the source's luminosity and the distance between the source and the observer, so Eq. 5.11 provides just what you need:

$$\text{apparent brightness} = \frac{L}{4\pi (dist)^2} = \frac{4 \times 10^{26} \text{ W}}{4\pi (1.5 \times 10^{11} \text{ m})^2}$$
$$= 1415 \text{ W/m}^2.$$

This is called the "solar constant" at Earth.

The following exercise will help you understand how tiny the apparent brightness from a typical (bright) star is.

Exercise 5.4. What is the apparent brightness at Earth of the star Vega, which has luminosity about 40 times that of the Sun and distance of approximately 25 light years from Earth?

5.3 Magnitudes

In addition to the quantities of apparent brightness and luminosity, astronomers have another system of classifying the brightness of celestial objects. That system is called the "magnitude" system, and it's based on the way the human eye sees different intensities of light. In this section, we'll explain the two main types of magnitude. "Apparent magnitude" is related to the apparent brightness of a light source, or how bright that source appears to an observer on Earth. "Absolute magnitude" is related to the intrinsic brightness, or luminosity of the source.

Table 5.1 *Hipparchus' magnitude system*

Apparent magnitude	Brightness
$m = 1$	Brightest stars in sky
$m = 2$	Bright stars
$m = 3$	Medium-brightness stars
$m = 4$	Dim stars
$m = 5$	Very dim stars
$m = 6$	Barely visible stars

5.3.1 Apparent magnitude

The apparent magnitude scale was introduced over 2,000 years ago by the Greek astronomer Hipparchus, who grouped the stars in the night sky into six categories. He lumped the 20 or so brightest stars he could see into the category called "first magnitude"; slightly dimmer stars went into the category called "second magnitude," still-dimmer stars into the category called 'third magnitude," and so on, to the very dimmest stars (the ones he could just barely see with his unaided eyes), which he put into the category called "sixth magnitude." So Hipparchus' system is basically a broad-category ranking of stars by their apparent brightness as seen from Earth. The apparent magnitude of a celestial object is often denoted as "m." Take care not to confuse this with "m," a variable often used to represent mass, and "m," the abbreviation for the length unit meters. The context should make it clear which of the three possible meanings of lower-case "m" is intended.

To give you a sense of Hipparchus' system, the bright stars in constellations such as Orion or the Southern Cross would have fallen into Hipparchus' first-magnitude group, most of the stars in the Big Dipper would belong to the second-magnitude group, and the stars of the Little Dipper would span several of Hipparchus' groups, from second to fourth or fifth magnitude. In today's light-polluted skies, stars in Hipparchus' sixth-magnitude category are invisible from all but the very best observing sites.

When scientists began developing techniques to make quantitative measurements of the brightness of light sources, they discovered an interesting fact: human vision and brightness perception operate in a logarithmic rather than linear fashion. That means that what we perceive as a certain *difference* in brightness (comparison by subtraction) is actually a *ratio* of brightness (comparison by division). To understand this, imagine that an observer is looking at three stars such as those shown in Figure 5.3: one very bright (say in Hipparchus' first-magnitude category), one less bright (perhaps in the second-magnitude category, and one still less bright (in the third-magnitude

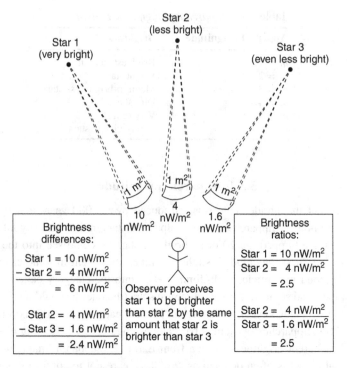

Figure 5.3 Comparing the brightness of three stars.

category). The observer is likely to say that the brightness difference between the first- and second-magnitude stars is the same as the difference between the second- and third-magnitude stars. But measurements reveal that in fact the second-magnitude stars are about 2.5 times dimmer than the first-magnitude stars, and the third-magnitude stars are about 2.5 times dimmer than the second-magnitude stars. So the *ratio* of brightness between categories is constant, but the *difference* in brightness is not.

If that seems contradictory, it may help to put some numbers into the system. Imagine the three stars seen by the observer have the following brightnesses: the brightest star (call it "star 1") has an apparent brightness of 10 nW/m^2 (that's nanowatts per square meter), the middle-brightness star ("star 2") has an apparent brightness of 4 nW/m^2, and the least-bright of the three stars ("star 3") has an apparent brightness of 1.6 nW/m^2. A human observer will perceive the brightness difference between star 1 and star 2 to be the same as the difference in brightness between star 2 and star 3. But the actual difference in apparent brightness between stars 1 and 2 is $10 - 4 = 6$ nW/m^2, while the difference between stars 2 and 3 is less than half that at $4 - 1.6 = 2.4$ nW/m^2. What *is*

the same between these pairs of stars is the ratio of their brightnesses, because
$\frac{10\,\text{nW/m}^2}{4\,\text{nW/m}^2} = 2.5$ and $\frac{4\,\text{nW/m}^2}{1.6\,\text{nW/m}^2} = 2.5$.

It's important to realize that a brightness ratio of approximately 2.5 for each step in magnitude means that the brightness ratio between a first-magnitude star and a third-magnitude star is approximately 6.3 (since $2.5 \times 2.5 \approx 6.3$), and the brightness ratio between a first-magnitude star and a fourth-magnitude star is about 16 (since $2.5^3 \approx 16$). Don't make the mistake of *adding* the factors of 2.5 for each magnitude step; you must *multiply* the brightness by 2.5 for each step in magnitude, or *divide* by 2.5 if you're counting toward dimmer stars.

In the nineteenth century British astronomer Norman Robert Pogson injected some specificity into the magnitude categories by suggesting that five steps in magnitude should correspond exactly to a factor of 100 in brightness. Thus the brightness ratio corresponding to one magnitude step should be $\sqrt[5]{100} = 2.512$. Figure 5.4 contains a table that illustrates the relationship between magnitude steps and brightness ratios.

This relationship can be put into an equation like this:

$$\text{brightness ratio} = (2.512)^{\#\ \text{mag steps}}, \tag{5.12}$$

where "# mag steps" (sometimes called Δm) represents the number of magnitude steps between the two objects.

Example: The apparent magnitude of the star Altair in the constellation of Aquila the Eagle is $m = 0.77$, and the apparent magnitude of the star Merak in the Big Dipper is $m = 2.36$. Which of these two stars is brighter, and by what factor?

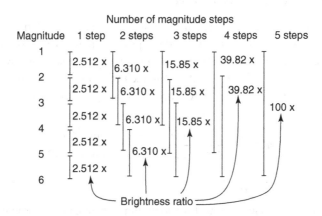

Figure 5.4 Magnitude steps and brightness ratios.

Since stars with higher positive magnitude numbers are dimmer, Merak is dimmer than Altair. And since each step in magnitude represents a factor of approximately 2.5 in brightness, you know that a difference of 1.59 magnitude steps ($\Delta m = 2.36 - 0.77 = 1.59$) means that Altair must be more than 2.5 times brighter than Merak. To get an exact number, you can use Eq. 5.12:

$$\text{brightness ratio} = (2.512)^{\Delta m} = (2.512)^{1.59}$$
$$= 4.325,$$

so Altair is a little over four times brighter than Merak.

Getting from number of magnitude steps to brightness ratio is straightforward using Eq. 5.12. But what if you're given a brightness ratio and you want to figure out how many magnitude steps that represents? To do that, you need to solve Eq. 5.12 for the number of magnitude steps, and the use of logarithms is very helpful for that.

Start by taking the logarithm (base 10) of both sides of Eq. 5.12:

$$\log_{10}(\text{brightness ratio}) = \log_{10}\left[(2.512)^{\Delta m}\right]. \tag{5.13}$$

That may not look helpful, but remember that $\log(x^a) = a\log(x)$. Applying this to the right side of Eq. 5.13 gives

$$\log_{10}(\text{brightness ratio}) = \Delta m \left[\log_{10}(2.512)\right]$$

or

$$\Delta m = \frac{\log_{10}(\text{brightness ratio})}{\log_{10}(2.512)}. \tag{5.14}$$

Example: If a certain star is 100 times dimmer than a star with m = 1, what is the apparent magnitude of the dimmer star?

You can do this one in your head if you remember that the modern magnitude scale is designed so that five magnitude steps correspond to a factor of 100 in brightness. So a star that's 100 times dimmer than a star with apparent magnitude of 1 must have an apparent magnitude that is five steps dimmer, which is apparent magnitude of 6. Here's how you can get that same result using Eq. 5.14:

$$\Delta m = \frac{\log_{10}(\text{brightness ratio})}{\log_{10}(2.512)} = \frac{\log_{10}(100)}{\log_{10}(2.512)}$$
$$= \frac{2}{0.4} = 5.0.$$

When you use a calculator to take logarithms in magnitude problems, it's important that you push the "log" button (which gives the desired base-10 logarithm) rather than the "ln" button (which gives the base-e or "natural" logarithm of the number).

Since optical instruments allow us to see objects that are too faint to be seen with the unaided eye, astronomers have extended the apparent magnitude system beyond the six categories of Hipparchus. So a star that is 2.512 times dimmer than a star of $m = +6$ is assigned $m = +7$, and a star that is 2.512 times dimmer than that star is assigned $m = +8$, and so on. Using a typical pair of binoculars under good skies, you can see stars down to m of around +10, and long-exposure photographs taken with the world's most-sensitive telescopes working with visible light have shown stars with m of around +30.

The reason we explicitly included "+" signs in front of the apparent magnitudes in the previous paragraph is that astronomers have also extended Hipparchus' magnitude scale to include brighter objects. And how shall we classify a star that's 2.512 times brighter than a star with $m = +1$? It must be a star of $m = 0$. And a star that's one magnitude step brighter than a 0-magnitude star must have $m = -1$. Continuing in that direction, astronomers assign $m = -4.9$ to the planet Venus at its brightest, and the full Moon has an m of about -12.8. The Sun is about 400,000 times brighter than the full Moon, which means that m for the Sun is about -26.8. You can verify this in the following exercise.

Exercise 5.5. Use the facts that the Sun appears about 400,000 times brighter than the full Moon and that the full Moon has $m = -12.8$ to verify that m for the Sun is about -26.8.

Some astronomy texts specify magnitudes as measured through a standard color filter by putting a subscript after the "m," such as m_B for apparent magnitude in the blue portion of the visible range only. If you encounter this notation, you can use Eqs. 5.12 through 5.14 in the same way you would use the apparent magnitude m – just recognize that the subscripted magnitude is referring to the object's brightness in one color only, and objects typically have different magnitudes in different colors.

5.3.2 Absolute magnitude

Once you're comfortable with the workings of the apparent magnitude scale, it's a small step to understanding absolute magnitude (often written as "M"). That's because the absolute magnitude of a celestial object is just the apparent magnitude the object *would* have if it were at a distance of 10 parsecs.

You can appreciate the value of absolute magnitude by considering the values of the apparent magnitude of a few familiar celestial objects.

A run-of-the-mill star in the night sky may have an m of around +3, and yet
the Sun (which is a run-of-the-mill star by most measures) has an m of −26.8.
That gigantic difference in apparent magnitude is due, of course, to the Sun's
proximity to Earth. The mere 93 million miles between the Earth and the Sun is
tens of thousands of times less than the distance to even the closest stars in the
night sky. And since brightness decreases as the square of distance, the Sun has
an advantage of several billion times in apparent brightness. So even though
the Sun's intrinsic brightness (its luminosity) is not especially large, its apparent magnitude differs by several dozen steps from the apparent magnitude of
other stars.

Absolute magnitude "levels the playing field" by considering not how bright
an object appears from Earth, but how bright that object would appear if it
were 10 pc away. By conceptually placing all objects at the same distance, the
$\frac{1}{(dist)^2}$ term that relates apparent brightness to luminosity is the same for all
objects, so absolute magnitude depends only on the luminosity of the object.
Thus you can be sure that stars that have small or negative absolute magnitudes
really are more luminous than stars that have large positive values of absolute
magnitude.

On the absolute-magnitude scale, the Sun comes in at about +4.8, as you can
see in the following example.

*Example: At a distance of 1 AU from the Earth, the Sun has $m = −26.8$. What
is the Sun's absolute magnitude?*

Problems like this beg for you to use the ratio method, because magnitude steps
are related to brightness *ratios*. Based on the definition of absolute magnitude,
you know that the Sun's absolute magnitude is the apparent magnitude that
the Sun would have if its distance from Earth were 10 pc. So a good approach
to this problem is to determine how many times farther away the Sun would
be if its distance were 10 pc, then to determine how many times dimmer it
would appear, and then to convert that many times dimmer into a number of
magnitude steps. You can then add that number of steps to the Sun's apparent
magnitude of −26.8.

To determine how many times farther the Sun would be if it were at 10 pc
instead of at 1 AU, you can make a ratio of these distances. But first you have
to make sure to put these two distances into the same units. To do that, you can
convert 10 pc to AU, or you can convert 1 AU to parsecs. We'll take the first
approach:

$$10 \text{ pc} = 10 \text{ pc} \times \left[\frac{206{,}265 \text{ AU}}{1 \text{ pc}} \right] = 2.06 \times 10^6 \text{ AU,}$$

which means the distance ratio is

$$\text{distance ratio} = \frac{10 \text{ pc}}{1 \text{ AU}} = \frac{2.06 \times 10^6 \text{ AU}}{1 \text{ AU}} = 2.06 \times 10^6.$$

So the Sun's distance would be about two million times greater if it were at 10 pc instead of 1 AU. Before you can turn that distance ratio into a number of magnitude steps, you first have to figure out how many times dimmer the Sun would appear if it were 2.06 million times farther away. But that's exactly what the inverse-square law tells you: if you move something twice as far away, it appears four times dimmer (that is, its brightness becomes one-quarter of its value at the closer position). The key is to *square* the distance ratio and put it in the denominator to get the brightness ratio, because brightness is proportional to $\frac{1}{(dist)^2}$. So, if the Sun were 2.06×10^6 times farther away, its brightness would decrease by a factor of $\frac{1}{(2.06 \times 10^6)^2} = \frac{1}{4.25 \times 10^{12}}$.

Thus the Sun would be about 4.25 trillion times dimmer if it were at a distance of 10 pc instead of its actual distance of 1 AU. To determine how many magnitude steps correspond to that many times dimmer, you can use Eq. 5.14:

$$\Delta m = \frac{\log_{10}(\text{brightness ratio})}{\log_{10}(2.512)} = \frac{\log_{10}(4.25 \times 10^{12})}{\log_{10}(2.512)}$$

$$= \frac{12.63}{0.4} = 31.6.$$

So the Sun's apparent magnitude would change by 31.6 steps if its distance from Earth were 10 pc. Since the Sun would be dimmer by this many steps, you have to add 31.6 to the Sun's apparent magnitude of -26.8 to get its absolute magnitude:

$$M_{Sun} = -26.8 + 31.6 = +4.8, \tag{5.15}$$

as expected.

And just as you may encounter subscripted apparent magnitudes such as m_B, you may also come across subscripted absolute magnitudes, which refer to the absolute magnitude in only one portion of the spectrum (so M_B refers to the absolute magnitude in the blue portion of the spectrum).

Exercise 5.6. Use the logic of the previous example to determine the absolute magnitude of a star that has $m = 3.2$ and distance of 175 ly.

5.3.3 Distance modulus

Just as you can find the distance to an object if you know its luminosity and apparent brightness, you can also find distance using absolute and apparent magnitude.

You can understand this by doing the following thought experiment: if a star's absolute magnitude is the same as its apparent magnitude, what can you conclude about that star's distance? Since the star's absolute magnitude is the apparent magnitude the star would have if its distance were 10 pc, and the star in this case actually does have that same apparent magnitude, it must be at a distance of 10 pc.

Now consider what it means if an object's apparent magnitude is greater (that is, a bigger positive number) than its absolute magnitude. Since a bigger positive magnitude means dimmer, at its actual distance the object appears dimmer than it would if its distance were 10 pc. And if it appears dimmer than it would at 10 pc, then its actual distance must be farther than 10 pc.

Likewise, if an object's apparent magnitude is a smaller positive number than its absolute magnitude, then the object appears brighter than it would if its distance were 10 pc. That can only mean that the object's actual distance is closer than 10 pc.

Knowing whether an object is closer or farther than 10 pc is useful, but you can determine precise distances by using the numerical difference between the object's apparent magnitude (m) and its absolute magnitude (M). The difference between these quantities is called the "distance modulus" (DM), and it's related to distance (d, measured in units of parsecs) by this equation:

$$DM = m - M = 5\log_{10}\left[\frac{d}{10\,\text{pc}}\right] \qquad (5.16)$$

which can be solved for distance:

$$10^{\frac{DM}{5}} = 10^{\log(d/10\,\text{pc})}$$

$$10^{\frac{DM}{5}} = \frac{d}{10\,\text{pc}}$$

$$d = 10\,\text{pc} \times \left[10^{\frac{DM}{5}}\right] = 10\,\text{pc} \times \left[10^{\frac{m-M}{5}}\right]. \qquad (5.17)$$

Example: The star Denebola in the constellation Leo is known by parallax measurments to lie about 36 light years from Earth, and Denebola's apparent magnitude is 2.1. What is Denebola's absolute magnitude?

Since you're given Denebola's apparent magnitude (m) and distance (d), you can use Eq. 5.16 to find Denebola's absolute magnitude (M). Begin by solving Eq. 5.16 for M:

$$DM = m - M = 5\log\left[\frac{d}{10\,\text{pc}}\right],$$

$$M = m - 5\log\left[\frac{d}{10\,\text{pc}}\right]. \qquad (5.18)$$

Since the distance in this equation must be in units of parsecs, before plugging in it's necessary to convert 36 light years to parsecs:

$$36 \text{ ly} = 36 \text{ ly} \times \left[\frac{1 \text{ pc}}{3.26 \text{ ly}} \right] = 11.0 \text{ pc},$$

which can be substituted into Eq. 5.18:

$$M = m - 5 \log \left[\frac{d}{10 \text{ pc}} \right] = 2.1 - 5 \log \left[\frac{11.0 \text{ pc}}{10 \text{ pc}} \right]$$
$$= 2.1 - 5 \log (1.1) = 1.9.$$

This is just a bit smaller than Denebola's apparent magnitude, as expected for a star that's just a bit farther than 10 pc from Earth.

Example: A variable star in the Andromeda Galaxy is measured to have an apparent magnitude of +18.5 and an absolute magnitude (which can be determined from its period of variation) of −6. How far away is the Andromeda Galaxy?

You're given both the apparent and the absolute magnitude, so the distance modulus may be found as

$$DM = m - M = 18.5 - (-6) = 24.5$$

from which the distance (d, measured in parsecs) may be found using Eq. 5.17:

$$d = 10 \text{ pc} \times \left[10^{\frac{DM}{5}} \right] = 10 \text{ pc} \times \left[10^{\frac{24.5}{5}} \right]$$
$$= 10 \text{ pc} \times \left[10^{4.9} \right] = 10 \text{ pc} \times (79{,}432.8) = 794{,}328 \text{ pc},$$

which is over 2.5 million light years. It makes sense that the apparent magnitude will be much larger (fainter) than the absolute magnitude for an object so far away. It also makes sense that the absolute magnitude is so bright (negative), because only a tremendously luminous star could be seen as far away as another galaxy.

Exercise 5.7. Use Eq. 5.18 to determine the absolute magnitude of a star that has $m = 3.2$ and distance of 175 light years.

5.4 H–R diagram

The Hertzsprung–Russell (H–R) diagram is a graph of two specific properties of many stars, but it organizes and summarizes many other properties of stars, so it is well worth your time to fully understand it. It is the most common

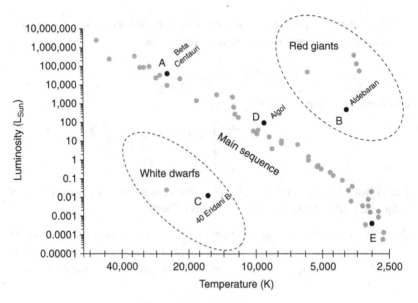

Figure 5.5 An H–R diagram.

graph in any study of astronomy that relates to stars; rare is the astronomy class or book that doesn't include an H–R diagram. It is also incredibly useful in revealing patterns of many different properties of stars in one simple graph, and the purpose of this section is to help you understand and interpret the graph. Graph reading is an important mathematical skill, so although you won't be performing many numerical calculations in this section, you will still be engaged in *quantitative reasoning*, which is every bit as important as being able to carry out arithmetic and algebraic operations.

Each dot on an H–R diagram, such as Figure 5.5, represents an individual star. The term "diagram" in other contexts often refers to a drawing or picture, but the H–R diagram is actually a *graph*.

You should understand that the properties of most stars remain fixed for the vast majority of their lives, and therefore a star's location on the H–R diagram remains fixed for most of its life. For most stars, that location falls on a broad swath running from the upper left to the lower right portions of the diagram; this is called the "main sequence," and it encompasses most of the stars plotted on the H–R diagram shown in Figure 5.5. The few stars that are not on the main sequence generally occupy regions in either the upper right, called "red giants," or lower left, called "white dwarfs."

The two star properties that are directly graphed on the H–R diagram are luminosity (L) and temperature (T). However, there are three other stellar properties that can be determined from the H–R diagram, even though they do not appear on the x- and y-axes. Those properties are star radius (R), lifetime (ℓ), and mass (M). One reason that the H–R diagram is such a powerful tool in astronomy is because it represents all five of these fundamental stellar properties in one compact graph. Each of these properties is discussed in this section, but you should begin by making sure you have a solid understanding of the concept of solar units.

5.4.1 Solar units

If you read that a certain star has a mass of 6×10^{30} kg, what do you make of it? Is that a large mass or a small mass, as stars go? Many of the quantities you'll be dealing with in astronomy involve numbers so large as to be meaningless, even to professional astronomers. In order to express physical quantities more intuitively, astronomers employ a set of "solarcentric" units that use our Sun's properties as a reference. Since our Sun is an average star, it makes a convenient reference. The subscript $\odot$ is frequently used to indicate that a parameter pertains to the Sun, so "$M_\odot$" and "M_{Sun}" both refer to the mass of the Sun. The key to understanding solar units is to realize that the Sun's properties can be used as the base units in which the properties of other stars are expressed.

The three most-common units that reference the Sun's properties are solar luminosities ($L_\odot$ or L_{Sun}), solar radii ($R_\odot$ or R_{Sun}), and solar masses ($M_\odot$ or M_{Sun}). A quantity of one of each of these units is defined as the Sun's value of that quantity. For example, since the Sun's mass is approximately 2×10^{30} kg, that equivalence can serve as a conversion factor: $1\,M_\odot \leftrightarrow 2 \times 10^{30}$ kg, or "One solar mass is equivalent to 2×10^{30} kg." Converting to and from solar units is a straightforward unit conversion problem. For example, the hypothetical star referred to in the preceding paragraph, with mass 6×10^{30} kg, can alternatively be expressed as

$$M_{star} = 6 \times 10^{30} \text{ kg} = 6 \times 10^{30} \text{ kg} \left(\frac{1\,M_\odot}{2 \times 10^{30} \text{ kg}} \right) = 3\,M_\odot,$$

which means that this star has a mass three times the mass of the Sun.

If the use of solar units feels unfamiliar to you, the following analogy may help. You could choose to express the heights of other things in terms of your own height (which you might call h_{you}). This allows you to know instantly whether those things are taller or shorter than you, and by what factor. For example, an Olympic gymnast might have a height of $0.9h_{you}$, and a big-city skyscraper might have a height of $230h_{you}$. Using these units, it's instantly clear that the gymnast is a bit shorter than you and that the skyscraper is many times taller than you. For the skyscraper in particular, whose height is much different than yours, using these egocentric units gives you a much better sense of its height than saying it is 380,000 mm tall. Similarly, expressing the mass of a hypothetical star as $3.0\,M_\odot$ instead of 6×10^{30} kg makes the quantity more intuitively meaningful. Expressed in solar units, you can instantly recognize that this is a star with mass several times greater than the mass of the Sun.

Here are the equivalence relations for the most-common solar units:

1 solar mass $\leftrightarrow 2 \times 10^{30}$ kg,
1 solar luminosity $\leftrightarrow 4 \times 10^{26}$ W,
1 solar radius $\leftrightarrow 6.96 \times 10^5$ km.

Exercise 5.8. Express the following in solar units:

(a) **Mass of Earth ($M_{Earth} = 6 \times 10^{24}$ kg)**
(b) **Mass of a star ($M_{star} = 8 \times 10^{32}$ kg)**
(c) **Radius of Earth ($R_{Earth} = 6{,}378$ km)**
(d) **Radius of our Solar System (use 40 AU)**
(e) **Luminosity of a star emitting 7×10^{31} W.**

5.4.2 Luminosity and temperature axes

The axes of the H–R diagram represent two fundamental observable properties of stars: luminosity (how bright a star is) and surface temperature (how hot it is on the outside, which is the only part we can directly see). Hereafter, star "temperature" will refer to "surface temperature" unless otherwise specified. Luminosity (with units of watts or $L_\odot$) is graphed along the y-axis in the standard fashion, with the smallest values at the bottom and largest values at the top.

Temperature is represented on the x-axis, in the standard SI units of kelvins (K). The most common form of the H–R diagram uses an unusual convention by representing the direction of increasing temperature values backwards compared to most graphs: temperature *increases* to the left and *decreases* to the right.

Example: Of the five stars labeled A through E on the graph in Figure 5.5, determine which is the brightest, which is the faintest, which is the hottest, and which is the coolest.

Since luminosity is graphed on the y-axis, the brightest star is the one with the largest y-value, closest to the top of the graph: point A. The faintest star, conversely, is the one with the smallest y-value, closest to the bottom of the graph: point E. Since temperature is graphed on the x-axis with the largest values to the left and smallest to the right, the hottest star is the one closest to the left side of the graph (point A), and the coolest star is farthest to the right (point E).

Notice in Figure 5.5 that the numerical values on the axes cover a huge range. Luminosity, in particular, ranges over more than 12 orders of magnitude.[1] In order to display this huge scale on one axis and avoid all the points bunching up at one end of the axis, a common practice in many fields of science is to use a "log scale" or spread the numbers logarithmically. This approach uses an even spacing for each multiplicative factor of some whole number (such as 10, 100, 1,000 or 2, 4, 8) instead of using evenly spaced consecutive numbers (such as 10, 11, 12). Ten is frequently chosen as the multiplicative factor (which makes the scale a "log 10 scale"); you can see this in practice on the luminosity (vertical) axis of Figure 5.5. On this axis, each major tick mark represents a number ten times larger than the previous (lower) tick mark: $\frac{1}{100,000}, \frac{1}{10,000}, \frac{1}{1,000}, \frac{1}{100}, \frac{1}{10}$, 1, 10, 100, ... up to 10,000,000. While 10 is the most commonly used factor on a log-scale axis, a multiplicative factor of 2 is sometimes used (making the scale a "log 2 scale"). You can see this in practice on the temperature (horizontal) axis of Figure 5.5, where each major tick mark represents a number two times larger than the previous (rightward) tick mark: 2,500, 5,000, 10,000, 20,000, 40,000. Remember that the H–R diagram x-axis has temperature increasing to the left.

Example: Read the luminosity and temperature values off the graph for each of the stars labeled in the H–R diagram in Figure 5.5.

This example is intended to help you get accustomed to reading logarithmic axes. For each point, imagine drawing a vertical and a horizontal line through it, and determine where these lines intersect the x- and y-axes. It may help to hold a straight edge horizontally or vertically through the point to see where it intersects the axes.

When reading between the labeled values on the axis, be sure to deduce the value of the intermediate tick marks by noting how many intervals there are and looking at the surrounding values. For example, on the x-axis between the temperature labels of 2,500 K and 5,000 K, there are five intervals (defined by four tick marks) spanning the range from 2,500 to 5,000 K, a range of 2,500 K.

---

[1] Be careful not to confuse "orders of magnitude" with the magnitude brightness scale discussed in Section 5.3. "Orders of magnitude" means "powers of ten."

Because the intervals have equal value, each represents 2,500 K ÷ 5 intervals = 500 K per interval. Note that the tick marks are not evenly spaced, since this is a logarithmic rather than a linear scale. So the star labeled "E," with a y-value one tick mark above 2,500 K, has a temperature of 3,000 K, and star "B" has a temperature of two ticks below 5,000 K, which is 4,000 K.

Note also that the intervals represent larger differences as you move up the temperature scale. For example, the next labeled range from 5,000 K to 10,000 K has five intervals again, but they now span 10,000 − 5,000 = 5,000 K, so each represents 5,000 K ÷ 5 intervals = 1,000 K per interval. This is twice as large as the intervals from the previous region. Similarly, the next region (from 10,000 K to 20,000 K) uses intervals of 2,000 K, which are twice as large again. Star "C" falls in this range, and its temperature is about two ticks below 20,000 K, which means its temperature is about 16,000 K. Deducing the intervals along the y-axis will work similarly, but you will find 9 intervals (8 ticks) in each range. So, for example, the tick marks between the values 1 and 10 represent values of 2, 3, 4, 5, 6, 7, 8, and 9 $L_\odot$. But between the values 100 and 1,000 the tick marks represent 200, 300, 400, 500, 600, 700, 800, and 900 $L_\odot$. Star "B" falls in this range, approximately on the fourth tick mark above 100, so its luminosity value is about 500 $L_\odot$.

A summary of all the parameters of the five named stars in Figure 5.5 is shown in Table 5.2.

Exercise 5.9. Determine the locations on the H–R diagram of stars with the following parameters:

Star 1: Temperature = 5,000 K, Luminosity = 0.3 L_{Sun}
Star 2: Temperature = 22,000 K, Luminosity = 0.02 L_{Sun}
Star 3: Temperature = 2,900 K, Luminosity = 400,000 L_{Sun}

5.4.3 Star radius

Although the radius of the star does not appear as one of the axes of the H–R diagram, it nonetheless varies in a very predictable manner on the graph.

Table 5.2 *Parameters of stars in Figure 5.5*

Star	Label	Temperature (K)	Luminosity ($L_\odot$)
Beta Centauri	A	25,000	40,000
Aldebaran	B	4,000	500
40 Eridani B	C	16,000	0.01
Algol	D	9,000	100
Groombridge	E	3,000	0.0004

The reason is that a star's radius is intimately related with its luminosity and temperature, which determine the star's position on the H–R diagram. Eq. 3.9 shows this relationship as $L \propto R^2 T^4$, which means that for any given temperature, stars with larger radius have higher luminosity. Likewise, for any given radius, stars with higher temperature have higher luminosity.

In order to see how radius varies in the H–R diagram, it is instructive to rearrange this relationship to solve for R^2:

$$R^2 \propto \frac{L}{T^4}. \qquad (5.19)$$

You could take the square root of both sides to get R to the first power ($R \propto \sqrt{L}/T^2$), but that is not necessary in order to understand how to estimate a star's radius from the H–R diagram.

From Eq. 5.19 above, you can see that R and L vary in the *same* sense, and R and T vary in an *inverse* sense – albeit to different powers. This means that if temperature is equal, larger L means larger R. That makes sense, because if two stars have the same temperature but one is more luminous than the other, the more-luminous star must be larger. Likewise, if two stars have the same luminosity but one is hotter than the other, the hotter star must be smaller.

What are the implications of this analysis for the H–R diagram? Think of it this way: to find stars with large radius R, look for stars with large luminosity and small temperature. Such stars appear in the upper-right corner of the diagram. Conversely, to find stars with small radius, look for stars with high temperature but small luminosity. These stars appear in the lower-left corner of the diagram. Regions with large and small radius are shown in Figure 5.6 This explains why the largest stars – red giants – are found in the upper-right quadrant of the diagram, and the smallest stars – white dwarfs – are found in the lower-left quadrant.

What about the upper-left and lower-right regions of the diagram? And the center, for that matter? How do the radii of those stars compare? Well, since those stars lie in the region between the extremes of small and large radius, they have intermediate radius. Such stars may have intermediate values of both L and T near the center of the H–R diagram, or large values of both L and T in the upper-left, or small values of both L and T in the lower-right portion of the diagram. Any star falling along the line connecting these regions will have roughly the same radius. This is why lines of constant radius run diagonally roughly from upper left to lower right on the H–R diagram, and are perpendicular to the direction of increasing radius. Several lines of constant radius are shown in Figure 5.6 as dashed lines.

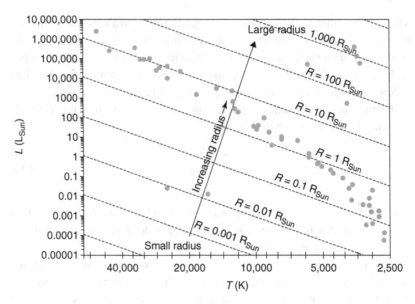

Figure 5.6 An H–R diagram with the direction of increasing radius.

Example: Rank the stars labeled in the H–R diagram in Figure 5.5 in order of increasing size.

Since radius increases from the lower left to the upper right, the smallest star is in the lower left: C. The largest star is in the upper right: B. Between those extremes, stars A, D, and E have similar radius. But if you look carefully at the lines of constant radius drawn in Figure 5.6, you will see that the lower end of the main sequence roughly coincides with the line of constant radius with $R = 0.1\,R_\odot$. Conversely, the upper end of the main sequence coincides with a constant-radius line of $R = 10\,R_\odot$, and the center of the main sequence is intermediate between the two, near the line for $R = 1\,R_\odot$. Hence star E has a smaller radius, star D is slightly larger, and star A is slightly larger still. So the ranking of all five from smallest to largest is C < E < D < A < B.

Exercise 5.10. Estimate the radius in kilometers of the five stars from the previous example.

5.4.4 Main-sequence star mass

The property that is most responsible for determining all other characteristics of main-sequence stars is their mass. Stars with more mass have stronger gravity, and therefore achieve higher core temperatures. Since fusion occurs only in the core, higher core temperatures produce higher fusion reaction rates, and this leads to higher luminosity. Moreover, inspection of Figures 5.5 and 5.6 show that luminosity and *surface* temperature are also correlated for main-sequence stars, because the highest-mass stars also have the highest surface temperatures. Thus, if mass is higher, the observable properties of temperature and luminosity are both higher as well. The relationship between mass and temperature cannot be given by a simple equation, but knowing that higher mass means higher temperature and luminosity allows you to apply quantitative reasoning, as the following example illustrates.

Example: Rank the labeled stars in Figure 5.5 by increasing mass.

Since the mass of main-sequence stars varies in the same sense as luminosity and temperature, you know that the star with the lowest T and smallest L (closest to the bottom right of the H–R diagram) must also have the lowest mass: star E. Similarly, the star with the highest T and largest L (closest to the top left of the diagram) must have the largest mass: star A. Star D is in-between. So the final ranking is E < D < A. You cannot include stars C and B in the ranking because they are not main-sequence stars, so they don't obey the relationship.

Some texts quote a quantitative relationship between main-sequence star mass and luminosity similar to the following:

$$L \propto M^{3.5}, \tag{5.20}$$

which indicates that stars with larger mass also have larger luminosity. Note that the exponent of 3.5 in Eq. 5.20 is only an approximation; the value of the exponent in the relationship varies between 3.0 and 4.0 for different regions on the main sequence. We will use 3.5 as a reasonable average for any main-sequence star, but be aware that other texts may use a slightly different value. The dependence of luminosity on mass is useful in the next section on estimating star lifetimes.

5.4.5 Main-sequence star lifetime

Stars do not live forever. They shine because of nuclear fusion reactions releasing energy[2] deep in their cores, but the fuel for those fusion reactions is finite and eventually runs out. The "lifetime" of a star is usually defined as the amount of time that star spends fusing hydrogen into helium in its core, which

---

[2] The amount of energy released by each nuclear reaction can be calculated by Einstein's famous equation $E = mc^2$, where m is the tiny amount of mass converted to energy in each reaction.

is the time the star spends on the main sequence. That time is determined by the amount of hydrogen fuel initially in the star's core (which is proportional to the star's mass) and the rate at which the star uses that fuel. Hence stellar lifetime problems fit the definition of "rate problems" discussed in Section 1.3.2. Here's an example that combines lifetime with fuel consumption rate:

Example: If a 5 $M_\odot$ star lives for 80 million years, what is its fuel consumption rate in kilograms per second?

The generic rate equation (Eq. 1.12) from Section 1.3 is

$$\text{amount} = \text{rate} \times \text{time}. \tag{1.12}$$

You are asked to calculate the rate of fuel use, so start by rearranging Eq. 1.12 to solve for rate:

$$\text{rate} = \frac{\text{amount}}{\text{time}}.$$

You can assume that the amount of fuel is directly proportional to the star mass, which is given as five times that of the Sun. The amount of the Sun's fuel was given as 9×10^{28} kg in the example from Section 1.3. Plugging in five times this quantity for amount, 80 million years for time, and including a conversion factor between years and seconds, will give you the rate in kilograms per second:

$$\text{rate} = \frac{5(9 \times 10^{28} \text{ kg})}{80 \times 10^6 \text{ yr}} \times \left(\frac{1 \text{ yr}}{3.1 \times 10^7 \text{ sec}} \right) = 1.8 \times 10^{14} \text{ kg/s},$$

which is over 300 times the Sun's rate.

Exercise 5.11. A certain high-mass star will fuse 2.9×10^{30} kg of hydrogen over its 10 million year lifetime. What is its rate of hydrogen use?

Exercise 5.12. Use the ratio method to calculate the lifetime of a star with 20 times more mass available for fusion than our Sun, but a fuel consumption rate 12,000 times faster than the Sun. For reference, the Sun's total lifetime is about 10 billion years.

Lifetime is another property that does not appear as an axis on the H–R diagram, but nonetheless varies smoothly and predictably in one direction along the main sequence. As stated above, stars with more mass have more fuel, because a portion of their mass *is* their fuel. While the entire mass of the star is not necessarily available to be used as fuel – only the hydrogen in the core is – the amount of available fuel is assumed to be directly proportional to the star's mass (this is a reasonable approximation for main-sequence stars).

So, while high-mass main-sequence stars do in fact have more fuel, they also have much higher luminosity as a result of their nuclear fusion occurring at a faster rate. This more prodigious fusion overcompensates for larger fuel reserves, burning through all the fuel much more quickly, resulting in a *shorter* lifetime for higher-mass main-sequence stars. Conversely, low-mass main-sequence stars have less fuel, but they meter it out much more parsimoniously, and as a result live for a longer time. Therefore the direction of increasing lifetime is *opposite* the direction of increasing mass on the main sequence. In other words, star lifetime and mass are *inversely* related, so as one gets larger the other gets smaller. The mathematical relationship between lifetime and mass can be seen by combining the assumption that a main-sequence star's available fuel is proportional to its mass (which can be stated as "amount of fuel $\propto M$") as discussed in the previous subsection, and the dependence of luminosity on mass as given by Eq. 5.20. Since the rate of fuel use is what directly determines the luminosity (power output), $L \propto M^{3.5}$ can be substituted in for "rate" in Eq. 1.12 as follows:

$$\text{lifetime} = \frac{\text{amount of fuel}}{\text{rate of fuel use}} \propto \frac{M}{L} \propto \frac{M}{M^{3.5}} \propto M^{1-3.5} \propto M^{-2.5}.$$

This mathematical relationship shows that mass and lifetime are not just inversely proportional: the inverse dependence of lifetime on mass is so strong that lifetime (ℓ) is inversely proportional to mass (M) raised to the power 2.5:

$$\ell \propto M^{-2.5} \quad \text{or} \quad \ell \propto \frac{1}{M^{2.5}} \quad \text{or} \quad \ell \propto \left(\frac{1}{M}\right)^{2.5}, \tag{5.21}$$

which are all mathematically identical statements. This means that a star that is half as massive as another does not just live twice as long, it lives $(\frac{1}{2})^{-2.5} = 5.7$ times as long as the more-massive star. As before with Eq. 5.20, the exponent of 2.5 in Eq. 5.21 is only an approximation. We will use 2.5 as a reasonable average for any main-sequence star, but other texts might use a slightly different value.

Example: What is the main-sequence lifetime of a 6 $M_\odot$ star?

Before doing any calculation, consider the fact that this star is *more massive* than the Sun (which has mass of 1 $M_\odot$, by definition). Therefore, since more-massive stars die sooner, you should expect that this star's lifetime will be less than the Sun's, which is about 10 billion years. So, before doing any math, you can predict that this star's lifetime is much less than 10^{10} years. You can get a quantitative answer by using Eq. 5.21 and applying the ratio method, using the Sun as the reference. Since this star's mass is a factor of six times the Sun's mass, this star's lifetime will be a factor of $6^{-2.5} = \frac{1}{6^{2.5}} = \frac{1}{88}$, or about 0.01 times the Sun's lifetime. That is, this star should live 0.01×10 billion years, or 100 million years. (Had you used a different value for the exponent, such as 2.0, your answer at this point would be $6^{-2.0} = \frac{1}{6^{2.0}} = \frac{1}{36} = 0.03$ times the Sun's lifetime, or 300 million years.)

Writing out the steps in the ratio method, the problem looks like this:

$$\frac{\ell_{star} \propto M_{star}^{-2.5}}{\ell_{Sun} \propto M_{Sun}^{-2.5}},$$

which becomes

$$\frac{\ell_{star}}{\ell_{Sun}} = \left(\frac{M_{star}}{M_{Sun}}\right)^{-2.5} = \left(\frac{6\,M_\odot}{1\,M_\odot}\right)^{-2.5} = (6)^{-2.5} = 0.01,$$

or

$$\ell_{star} = 0.01 \times \ell_{Sun} = 0.01 \times (10^{10} \text{ years}) = 100 \text{ million years.}$$

Exercise 5.13. What is the mass of a star that has a main-sequence lifetime of one billion years?

You may find it helpful to sketch a bare-bones H–R diagram with arrows indicating the direction in which each of the parameters discussed in this section increases. For luminosity and temperature, you can just sketch the usual H–R diagram axes, with L on the vertical axis, and an arrow pointing upward since larger luminosities are higher on the graph. The horizontal axis should be labeled with a T, with the arrow pointing to the left (toward the origin) since higher temperatures are farther to the left on the graph. A diagonal line extending up and to the right from the origin will be labeled R, since radius increases up and right on the graph. Both ℓ and M will be labeled along the main sequence, with M increasing up and to the left, and ℓ increasing down and to the right. Figure 5.7 shows arrows indicating the direction of increase of all five properties of stars discussed in this section.

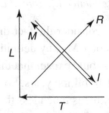

Figure 5.7 Star properties summarized on a schematic H–R diagram.

5.5 Chapter problems

5.1 Show how the general parallax equation (Eq. 4.3 of Chapter 4) leads to the stellar parallax equation (Eq. 5.1). Hint: solve Eq. 4.3 for distance, use a baseline of 2 AU, and convert to parsecs for distance and arcseconds for parallax angle. Then remember that the general parallax equation pertains to the full parallax angle while the stellar parallax equation uses *half* of that angle.

5.2 The distance to the Andromeda Galaxy from Earth is about 2.4 million light years. What is the parallax angular shift of Andromeda measured from Earth?

5.3 A distant galaxy is 3,000 times farther than Andromeda. What is the parallax angular shift for this galaxy?

5.4 How big would an optical telescope have to be in order to have resolution equal to the parallax angular shift of the galaxy of the previous problem?

5.5 The entire world consumes about 4×10^{20} joules of energy per year (primarily from fossil-fuel sources).

   (a) How long does it take the Sun to output that much energy?

   (b) What area of solar panels would be required on (or near) Earth to meet the Earth's energy demands solely with sunlight? Assume the sunlight is shining continuously straight down on all the solar panels and the solar power is captured with 100% efficiency.

   (c) How does the area of solar panels compare to the surface area of the Earth?

5.6 You determine the temperature of a certain star to be 8,000 K. Its apparent brightness is 4×10^{-10} W/m^2. Assuming it is on the main sequence, use the H–R diagram to estimate the distance to the star. Will its lifetime be longer or shorter than the Sun's?

5.7 A particular star has a parallax angle of 0.02 arcsec, an apparent brightness of 7×10^{-9} W/m^2, and its spectrum peaks at 500 nm. Calculate the luminosity and radius of the star in solar units.

5.8 If two stars have the same temperature but one is 10 million times more luminous, how do their radii compare?

5.9 Near the end of its life, our Sun's radius will increase to 1 AU, while its temperature will decrease to 3,500 K. What will happen to its luminosity?

5.10 Find the parallax angular shift of a star with apparent magnitude of $+10.1$ and absolute magnitude of $+3.2$.

6

Black holes and cosmology

Two of the most popular topics in astronomy classes are black holes and cosmology. Both of these subjects can be somewhat abstract, hard to visualize, and quite mathematical, giving them a mystique which likely contributes to their popularity. Precisely because some of the objects and processes are hard to visualize, the mathematical foundations of these topics are a valuable source of insight into their nature. So for these topics, even more than for other topics for which you have physical intuition on your side, it behooves you to understand the mathematics.

This chapter deals with "limiting cases" by investigating the mathematical ramifications of taking one variable to an extreme, such as allowing the radius of an object to shrink to zero or permitting time to run to infinity. The physical manifestations of these mathematical limiting cases lead to the most exotic concepts in astronomy: black holes, which are singularities of mass; and cosmology, which deals with the history and fate of the Universe. The chapter draws upon many of the tools discussed previously in this book, including units, solving equations using ratios and the absolute method, gravity, light, and graph interpretation. Black holes and cosmology bookend the entire range of possible sizes, from the infinitesimally small to the unimaginably immense, and are well worth the investment of time it takes to understand their mathematical foundations.

Before diving into black holes, you should make sure you have a solid understanding of the concepts and equations related to density and escape speed. Those are the subjects of the first two sections of this chapter, so if you're already comfortable with those topics, you can jump ahead to Section 6.3.

6.1 Density

Many students recognize the word "density" but are uncertain about its meaning. A common misconception is that density is the same as weight or mass. As you'll see in this section, mass and density are related, but they are not the same. That's because density is also related to another physical property you've already seen in this book: volume, which is the amount of space occupied by an object. In addition to providing the definition of density, this section will also show you how to calculate density for simple geometrical shapes and how to analyze the limiting cases in which density approaches zero and infinity.

6.1.1 Definition of density

Consider three small cubes, each 1 cm on a side. Imagine they are all painted grey so they all look the same, but one is composed of styrofoam, one of wood, and one of rock.

Imagine holding each of these cubes in your hand or putting them on a scale. Which one would you expect to weigh the most? You probably have an intuitive sense that the rock cube would weigh the most and the styrofoam cube the least. And since weight is a measure of the force of gravity on an object (see Section 2.1) and that force depends on the object's mass, the three objects must have different masses. In this case, the rank from smallest to largest mass is $m_{styrofoam} < m_{wood} < m_{rock}$. Remember that the SI unit for mass is kilograms (kg); other mass units used in astronomy are grams (g) and solar masses ($M_{\odot}$).

As mentioned above, the volume of an object is the amount of physical space occupied by the object. For a simple geometrical object like a rectangular solid or "parallelepiped," of which a cube is a special case, volume (V) can be calculated as length × width × height. Since all three cubes in Figure 6.1 have the same dimensions, all three cubes have identical volumes: $V_{styrofoam} = V_{wood} = V_{rock}$. In three-dimensional space, the dimensions of

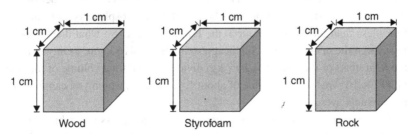

Figure 6.1 Three cubes of different material with identical appearance and dimensions.

volume are length to the third power, with SI units of cubic meters (m^3); other volume units you may encounter are cubic centimeters (cm^3) and cubic parsecs (pc^3).

So the three cubes have different masses but identical volumes. What does this have to do with density? Density is a measurement of how closely packed together the matter in an object is. Thus density relates to how heavy an object is relative to its size. If an object is "light" (small mass) and fluffy (large volume), it has a low density. If it is "heavy" (high mass) and compact (small volume), it has a high density. The reason "light" and "heavy" are in quotes in the previous sentence is that density is more than a measurement of mass or weight; it also depends on size. You can imagine a mountain made of cotton balls that has a very low density but is still far too heavy for you to lift, or a lead paperweight that is very dense but which you could lift easily. A crucial part of the definition of density is alluded to by the qualifier "relative to its size." Thus density depends on both the mass and the volume of the object, as you can see in the following definition:

$$\text{density} = \frac{\text{mass}}{\text{volume}}. \tag{6.1}$$

This equation makes it clear that the dimensions of density are the dimensions of mass divided by the dimensions of volume, which means the SI units of density are kilograms over cubic meters (kg/m^3). In astronomy, you may also see densities expressed in units of grams per cubic centimeter (g/cm^3).

Density is usually specified as a characteristic of a material rather than an entire object, since an object may be made of multiple materials of different densities. Even if an object is made of only one material, the density may be different at different locations within the object due to pressure and temperature differences within the object. For example, deep within a star the weight of overlying layers produces tremendous force that compresses the material in the core, which means that density increases with depth. Within the Sun, the density of hydrogen in the core is dozens of times greater than the density of iron, but the density of hydrogen at the top of the photosphere is thousands of times less than the density of air at the Earth's surface. So what does it mean to say that the density of the Sun is approximately 1,416 kg/m^3? Only that the the total mass of the Sun (2×10^{30} kg) divided by the total volume of the Sun (1.412×10^{27} m^3) gives a value of about 1,416 kg/m^3, which you can think of as the Sun's average density.

Some materials (including most liquids) are incompressible, which means that their density does not depend on the pressure exerted upon them. Additionally, some whole objects are homogeneous, which means that they are uniform throughout their volume. In such cases, a measurement of the mass and volume of any small sample can be used to determine the density of the material; the size or location of the sample does not matter, since the density must be the same everywhere within the object. For example, the density of seawater is everywhere about 1,000 kg/m^3, but you don't need to know the mass and volume of an entire ocean in order to measure that.

Example: Rank the cubes in Figure 6.1 in order of increasing density.

You know from the earlier discussion the order of the cubes by mass is $m_{styrofoam} < m_{wood} < m_{rock}$. Since density = mass/volume and all three cubes have the same volume, their order by density will be the same as their order by mass: density$_{styrofoam}$ < density$_{wood}$ < density$_{rock}$.

One way to estimate the density of an object is to imagine placing the object in water. If you dropped all three cubes in water, the styrofoam cube would float very high on the water, the rock cube would sink rapidly, and the wooden cube would be intermediate, probably floating low in the water. The reason that these identically sized cubes find different equilibrium points in the water is that they have different densities. Styrofoam ($\sim$75 kg/m^3) has a much lower density than water (1,000 kg/m^3), wood ($\sim$700 kg/m^3) has a density only slightly lower than water, and rock ($\sim$3,000 kg/m^3) has a density much higher than water.

Example: What is the density of a material for which a cube with sides of 1 cm has a mass of 0.7 g?

Since you're given the size and mass of the object and you're trying to find the density, you can use Eq. 6.1 to solve this problem. The first step is to find the volume of the cube:

$$\text{volume} = \text{length} \times \text{width} \times \text{height} = (1 \text{ cm}) \times (1 \text{ cm}) \times (1 \text{ cm}) = 1 \text{ cm}^3,$$

which converts to cubic meters as

$$\text{volume} = 1 \text{ cm}^3 \times \left(\frac{1 \text{ m}}{10^2 \text{ cm}}\right)^3 = 1 \text{ cm}^3 \times \frac{1 \text{ m}^3}{10^6 \text{ cm}^3} = 1 \times 10^{-6} \text{ m}^3.$$

Plugging this volume and the mass given in the problem statement (0.7 g or 0.0007 kg) into Eq. 6.1 gives

$$\text{density} = \frac{\text{mass}}{\text{volume}} = \frac{0.0007 \text{ kg}}{1 \times 10^{-6} \text{ m}^3}$$
$$= 700 \text{ kg/m}^3,$$

which is approximately the density of wood.

Exercise 6.1. Given the densities of styrofoam and rock in this section, calculate the masses of 1-cm cubes of each material.

6.1.2 Density proportionalities and limiting cases

It is often instructive to consider what would happen if you let one of the physical parameters in an equation go to a limiting value such as zero or infinity. For example, consider the limiting cases of the values of mass and volume on the right side of Eq. 6.1. The possibilities can be explored with a few mathematical thought experiments.

First, imagine keeping the volume fixed and letting the mass vary. Since density = mass/volume, mathematically this means making the numerator on the right side bigger or smaller while keeping the denominator the same, and seeing how the left side changes as a result. Visualize a sphere of fixed volume, say the size of an orange. If you gave it less and less mass, by first making it out of lead, then rock, then wood, then styrofoam, then air, approaching empty space, then the density would approach zero. Conversely, if you kept making the sphere more and more massive at fixed volume, the density would approach infinity. This is because with volume held constant, density and mass are directly proportional: density $\propto$ mass.

Now imagine holding the mass constant while varying the volume. That is, you have a fixed amount of "stuff" (so the mass remains constant), but you can spread it out over more space or compress it into less space. Mathematically, you'd be changing the denominator of the right side of the density equation (density = mass/volume) while keeping the numerator the same, and seeing how the left side changes. A loaf of spongy bread works well for a physical analogy. If you could spread that material out into a bigger volume, it would become fluffier, with more air or empty space between the bread particles. Eventually the material would become so spread out as to become imperceptible; as volume approached infinity, the density would approach zero. Conversely, if you compressed the bread, it would become a hard, compact little nugget with high density. The more compact, the smaller the volume, the higher the density. As volume approached zero, the density would approach infinity. This is because with mass held constant, density and volume are inversely proportional: density $\propto \frac{1}{V}$. In the limiting mathematical cases, $0 = \frac{1}{\infty}$ and $\infty = \frac{1}{0}$ for large and small volume, respectively.

As an astronomical example, consider a star at the end of its life. Some massive stars explode as supernovae at the end of their lives and disperse most of their mass into space. That ejected material never disappears entirely, but eventually gets so spread out that it becomes undetectable as it blends into the diffuse gas that fills the near-vacuum of interstellar space. The fixed amount of ejected mass gets spread over an ever-increasing volume ($\frac{m_{star}}{V \to \infty}$), so its density approaches zero.

Conversely, the cores of some stars collapse at the end of their lives under the inexorable crush of gravity when there is no longer any fusion to support them. Some actually collapse down to a volume of zero, becoming a "singularity," which is an infinitesimally small point occupying no physical space. But since they retain all the mass that went into them ($\frac{m_{star}}{V \to 0}$), the matter approaches infinite density. Such objects are called black holes, which are the subject of Section 6.3.

Exercise 6.2. How do the mass, volume, and average density of a car change when it is crushed by a compactor at the junk yard?

6.1.3 Density of spherical objects

Most of the astronomical objects that you are likely to encounter in density problems are roughly spherical in shape: planets, large moons, and stars are a few examples. Since the volume of a sphere is $V = \frac{4}{3}\pi R^3$ where R is the sphere's radius, the density of a sphere of mass m is

$$\text{density} = \frac{m}{\frac{4}{3}\pi R^3}. \tag{6.2}$$

Example: Use the radius and mass of the Sun to verify the average density of the Sun given in the previous section.

The radius and mass of the Sun are $1\ R_\odot = 6.96 \times 10^5$ km (or 6.96×10^8 m) and $1\ M_\odot = 1.99 \times 10^{30}$ kg, respectively. Plugging these values into Eq. 6.2 gives

$$\begin{aligned}
\text{density} &= \frac{m}{\frac{4}{3}\pi R^3} \\
&= \frac{1.99 \times 10^{30}\ \text{kg}}{\frac{4}{3}\pi(6.96 \times 10^8\ \text{m})^3} = \frac{1.99 \times 10^{30}\ \text{kg}}{\frac{4}{3}\pi(3.37 \times 10^{26}\ \text{m}^3)} \\
&= 1.41 \times 10^3\ \frac{\text{kg}}{\text{m}^3}, \text{ or about } 1{,}400\ \text{kg/m}^3.
\end{aligned} \tag{6.3}$$

Notice that density has dimensions of mass per unit volume, as expected. A two-step unit conversion will translate the units of the answer from kg/m³ into g/cm³, using the conversion factors $1\ \text{kg} \leftrightarrow 10^3$ g and $1\ \text{m} \leftrightarrow 10^2$ cm:

$$1{,}400\frac{\text{kg}}{\text{m}^3}\left(\frac{10^3\ \text{g}}{1\ \text{kg}}\right)\left(\frac{1\ \text{m}}{10^2\ \text{cm}}\right)^3 = 1{,}400\,(10^3\ \text{g})\left(\frac{1}{10^6\ \text{cm}^3}\right) = 1.4\ \text{g/cm}^3.$$

Example: How do the densities of the Sun and Earth compare? The Sun's radius is approximately 100 times that of Earth, and the Sun is about 300,000 times more massive than Earth.

Since this is a comparison problem, it is best solved using the ratio method from Section 1.2.3. To use this approach, begin by writing the equation for density for both objects:

$$\text{density}_{Sun} = \frac{\text{mass}_{Sun}}{\text{volume}_{Sun}} = \frac{m_{Sun}}{\frac{4}{3}\pi R_{Sun}^3}$$

and

$$\text{density}_{Earth} = \frac{\text{mass}_{Earth}}{\text{volume}_{Earth}} = \frac{m_{Earth}}{\frac{4}{3}\pi R_{Earth}^3}.$$

Now divide the Sun equation by the Earth equation:

$$\frac{\text{density}_{Sun}}{\text{density}_{Earth}} = \frac{\frac{m_{Sun}}{\frac{4}{3}\pi R_{Sun}^3}}{\frac{m_{Earth}}{\frac{4}{3}\pi R_{Earth}^3}} = \frac{\frac{m_{Sun}}{\frac{4}{3}\pi R_{Sun}^3}}{\frac{m_{Earth}}{\frac{4}{3}\pi R_{Earth}^3}} = \frac{m_{Sun}}{R_{Sun}^3}\frac{R_{Earth}^3}{m_{Earth}}$$

$$\frac{\text{density}_{Sun}}{\text{density}_{Earth}} = \frac{m_{Sun}}{m_{Earth}}\left(\frac{R_{Earth}}{R_{Sun}}\right)^3. \qquad (6.4)$$

The next step is to substitute in values from the comparison information given in the problem. Translating words into math, the problem gives you two key relationships to use as substitutions: "The Sun's radius is 100 times Earth's radius" becomes $R_{Sun} = 100 R_{Earth}$, and "The Sun's mass is 300,000 times the Earth's mass" becomes $m_{Sun} = 300,000 m_{Earth}$. Thus

$$\frac{\text{density}_{Sun}}{\text{density}_{Earth}} = \frac{300,000 m_{Earth}}{m_{Earth}}\left(\frac{R_{Earth}}{100 R_{Earth}}\right)^3$$

$$= \frac{300,000}{(100)^3} = \frac{3}{10} = 0.3.$$

This shows that the Sun's density is about one-third of the Earth's density. Don't forget the critically important step of asking yourself "Does my answer make sense?" In this case, the answer is "Yes," because the Sun is made of gas, and the Earth is predominantly rock, so it's reasonable that the Earth has a higher average density.

Exercise 6.3. The gas-giant planet Jupiter has a radius of 71,500 km and a mass of 1.9×10^{27} kg. The rocky planet Mercury has a radius of 2,440 km and a mass of 3.3×10^{23} kg. How do the densities of these two planets compare?

6.2 Escape speed

You may already have an intuitive sense for escape speed (sometimes called "escape velocity," and usually abbreviated v_{esc}).[1] Unlike many terms in astronomy, "escape speed" is self-descriptive: it is the speed required for one object to "escape" from another object despite its gravitational pull. That is, once the escaping object reaches escape speed, it will not fall back onto the other object even with no additional propulsion. You shouldn't let this lead you to the incorrect conclusion that the gravitational pull of an object somehow "turns off" at a certain distance and that to escape from it you simply need to reach that distance. Gravity never actually turns off, it just gets weaker and weaker with increasing distance, as you can tell from the R^2-term in the denominator of the equation for the force of gravity: $F_g = G\frac{m_1 m_2}{R^2}$. So unless you're infinitely far away from an object, its gravitational force on you will never be zero, which means you can never really "escape" from its gravitational pull. But if you're moving fast enough – specifically, at v_{esc} or faster – that gravitational pull will not be strong enough to cause you to fall back onto the other object.

6.2.1 Escape speed conceptual explanation

It is certainly possible to move away from a planet, moon, star, or other gravitating object without reaching escape speed – birds, balloons, and aircraft all leave the surface of the Earth without ever reaching escape speed. That's because each of these objects uses some form of propulsion or lift to oppose the downward force of Earth's gravity. But if those propulsion and lift forces are removed, Earth's gravity will cause these objects to fall back to the ground. In contrast, objects that have achieved escape speed need no additional force to continue moving away.

To understand how this works, consider the flight of a cannonball fired from the surface of the Earth at various speeds, as shown in Figure 6.2. In this figure, the only force acting on the projectile after it leaves the barrel of the cannon is the Earth's gravity, so the effect of air resistance is ignored.

As you can see in Figure 6.2(a), projectiles fired with speed less than the circular-orbit speed (called v_{circ}) are pulled to the ground after some distance by Earth's gravity. As common experience suggests, the faster the object is moving, the farther it goes before hitting the ground. But if the speed of the

---

[1] Many astronomy texts use these terms interchangeably, but we prefer "escape speed" since velocity is a vector (that is, it has a direction as well as a magnitude) and speed is a scalar, and no direction is implied in escape speed.

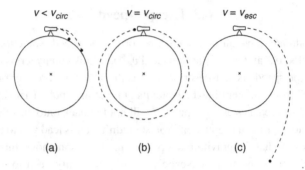

Figure 6.2 Projectiles fired at various speeds.

cannonball is equal to v_{circ}, the force of Earth's gravity produces an acceleration that curves the projectile's path just enough so that it makes a perfect circle around the center of the Earth, as shown in Figure 6.2(b). Remember that acceleration can mean speeding up, slowing down, or – as in this case – turning. This is how satellites in circular orbits remain above the Earth without the need for continuing propulsion.

If you're wondering if it would be possible to cause an object to orbit the Earth just a few meters above the surface, the answer would be "Yes" if the Earth were perfectly spherical (no mountains or other obstructions in the path) and had no atmosphere to slow the projectile down.

Now imagine firing the projectile much faster, with a speed equal to the escape speed v_{esc}. In that case, the force of Earth's gravity will still cause the path to curve, but that curve will be a parabola, as shown in Figure 6.2(c). Unlike a circle or an ellipse, a parabola is an open curve, meaning it does not close back on itself – so a projectile on a parabolic trajectory never returns to its starting point. This is an example of an "unbound" orbit.

You may be wondering what would happen to a projectile fired with speed slightly less or slightly more than v_{esc}. The answer is that an object with speed between v_{circ} and v_{esc} will follow an elongated but still bound elliptical orbit, and the path of any object with speed greater than v_{esc} will be an unbound hyperbola.[2]

You may also be wondering if the direction of the projectile matters in the determination of escape speed. The answer is "No" as long as the parabolic or hyperbolic path of the projectile doesn't intersect the surface of the object from which it is escaping. So whether you point your cannon horizontally, vertically,

---

[2] Circles, ellipses, parabolas and hyperbolas are all forms of "conic sections"; you can find links to additional information on the book's website.

or something in between (but not down at the ground), you'll never see your cannonball again if you fire it with speed of v_{esc} or greater.

It may help your understanding of escape speed to consider the case of an object launched vertically from the surface of the Earth. Imagine throwing an apple straight up in the air. Your arm imparts to the apple some initial velocity but, from the instant it leaves your hand, the only force acting on the apple is the downward force of gravity (again neglecting air resistance). Under the influence of this downward force, the apple will slow in its upward motion – remember from Section 2.2 that any unbalanced force produces acceleration in the direction of the force; when acceleration and velocity point in opposite directions, the object slows down. As long as the initial speed you give the apple is less than v_{esc}, the apple will eventually stop going up and fall back down. If you gave the apple a larger initial velocity by throwing it harder, it would go farther up before falling back down, but it would still fall back to the ground.

Now imagine giving the apple an initial speed equal to or greater than v_{esc}. In that case, the apple would still slow down under the influence of gravity, but it would never slow enough to stop, reverse, and fall back to the ground. Earth's gravity would still be acting on the apple forever, trying to pull it back to the Earth, and the apple would continue slowing down due to the force of Earth's gravity. But you know from Newton's Law of Gravity in Section 2.1 that the Earth's gravity gets weaker as the apple moves away, and with weaker force comes smaller acceleration, so the slowing of the apple would diminish. As time and the apple's distance from Earth approached infinity, the apple's speed would continue to slow down, but the apple would never quite stop, and certainly never turn around and come back. If the apple's initial speed were exactly equal to v_{esc}, the apple would just barely escape, meaning that it would never return to Earth but its speed would asymptotically approach zero as time approached infinity. If the initial speed were greater than v_{esc}, the apple's speed would never dwindle to zero, and the apple would escape with speed to spare.

6.2.2 Calculating escape speed

To calculate the value of v_{esc} at a given distance from the center of a given mass, use the equation

$$v_{esc} = \sqrt{\frac{2Gm}{R}} \tag{6.5}$$

in which G is the universal gravitational constant, m is the mass of the object from which the other object is escaping, and R is the distance between the centers of the objects. This relationship comes from an equivalence between the change in the kinetic energy (energy of motion) and the change in the gravitational potential energy (energy of position) of an object moving in a gravitational field. If you're interested in the details of how this relationship is derived, you can find more information about this on the book's website.

The placement of m and R on the right side of Eq. 6.5 has physical significance. Since m is in the numerator, v_{esc} gets larger as m gets larger. This is because objects with larger mass (m) exert greater gravitational force, and so it is harder to escape from them. Since the distance R is in the denominator, its inverse relationship means that v_{esc} gets smaller as R gets larger. This is because as you get farther from an object, the gravitational force from it gets weaker, so it is easier to escape from it.

Example: Calculate the escape speed at the surface of the Earth.

In order to apply Eq. 6.5, you need to know the Earth's mass (m) and the distance (R) from the center of that mass to the point at which you're calculating the escape speed. Since in this problem you're trying to find v_{esc} from the surface of the Earth, the distance between a hypothetical projectile and the center of the Earth is equal to Earth's radius. You can look up the mass and radius of the Earth (as well as the value of G) in any comprehensive astronomy text or on the Internet, where you should find that $G = 6.67 \times 10^{-11} \mathrm{N}\frac{m^2}{kg^2}$ (or $\frac{m^3}{kg\,s^2}$), $m_{Earth} = 6 \times 10^{24}$ kg, and $R_{Earth} = 6.4 \times 10^6$ m. Plugging these in gives

$$v_{esc,Earth} = \sqrt{\frac{2Gm_{Earth}}{R_{Earth}}}$$

$$= \sqrt{\frac{2(6.67 \times 10^{-11}\frac{m^3}{kg\,s^2})(6 \times 10^{24}\,kg)}{6.4 \times 10^6\,m}}$$

$$= \sqrt{1.25 \times 10^8 \frac{m^2}{s^2}}$$

$$= 1.1 \times 10^4 \text{ m/s, or 11 km/s.}$$

Notice that in this calculation the mass of the escaping object was not known and not needed. This means that the escape speed from the surface of the Earth is the same for all objects regardless of their mass, from an air molecule to a spaceship. If you find it surprising that escape speed doesn't depend on the mass of the object doing the escaping, remember that although a higher-mass projectile will indeed feel a stronger force of gravity than a lower-mass projectile, a higher-mass object also resists acceleration more than a lower-mass object since higher-mass objects have greater inertia. So, much like objects of all masses fall to the ground with the same acceleration in the absence of air resistance, objects of all masses have the same escape speed at a given distance from the Earth or any other gravitating object.

Example: If you could crush all the mass of the Earth into a smaller size – a sphere half its present radius – what would happen to the escape speed from its new surface?

Since you are asked to compare two scenarios, this is a good opportunity to use the ratio method. The problem specifies that the mass of the Earth remains the same ($m_{small\text{-}Earth} = m_{Earth}$), but the radius becomes half as large ($R_{small\text{-}Earth} = \frac{1}{2}R_{Earth}$). Writing Eq. 6.5 once for each object and then substituting in the equivalences just identified gives

$$\frac{v_{esc,\,small\text{-}Earth}}{v_{esc,\,Earth}} = \frac{\sqrt{\frac{2Gm_{small\text{-}Earth}}{R_{small\text{-}Earth}}}}{\sqrt{\frac{2Gm_{Earth}}{R_{Earth}}}} = \frac{\sqrt{\frac{2Gm_{Earth}}{\frac{1}{2}R_{Earth}}}}{\sqrt{\frac{2Gm_{Earth}}{R_{Earth}}}} = \frac{\sqrt{\frac{1}{\frac{1}{2}}}}{\sqrt{1}} = \frac{\sqrt{2}}{1} \simeq 1.4.$$

$$v_{esc,\,small\text{-}Earth} = 1.4\,v_{esc,\,Earth}.$$

So even if two different objects have the same mass, the escape speed from their surfaces will be different if the two objects have different radii. In particular, the more compressed the mass (the higher the density and thus smaller the radius), the larger the escape speed from it. Thus, for an object of given mass, v_{esc} from its surface will vary inversely as the object's density. This relationship will be important in the definition of black holes in the next section, but first here are a few exercises to give you a chance to practice using the equation for escape speed.

Exercise 6.4. How does the escape speed from low Earth orbit (where the space station orbits) compare to that from the surface? Assume an altitude of 350 km above the surface for low Earth orbit.

Exercise 6.5. Calculate the escape speed from the Sun at the distance of Earth's orbit.

Exercise 6.6. Now consider compressing the Sun into a black hole with the exact same mass (this could never happen in reality, because black holes don't form in nature with such low mass, but consider it hypothetically). What would be the escape speed from the "black-hole Sun" at the distance of Earth's orbit? Make sure you understand why this answer is either the same or different from your previous answer.

6.3 Black holes

In this section you will apply the concepts from the previous two sections –
density and escape speed – to some of the most exotic and bizarre objects in
the Universe: black holes. By definition, a black hole exists at any location at
which the density of matter is so great that the escape speed from the vicinity of
that location equals or exceeds the speed of light. Theoretically, any amount of
mass can be compressed to the density at which it becomes a black hole, but in
an introductory astronomy class, you are most likely to encounter black holes
of several solar masses that form when the cores of massive stars collapse, or
those with millions or billions of solar masses at the centers of galaxies.

6.3.1 Density of a black hole

The death of a high-mass star with a mass exceeding about 8 $M_\odot$ is an
extremely violent process in which the star's outer layers are blown off in a
supernova explosion, leaving the extremely dense core behind. If the mass of
that remnant core is greater than about 3 $M_\odot$, there's no force in the Universe that can prevent gravity from causing the material of the core to collapse
into a singularity. As the remnant core shrinks, its mass remains the same, but
its volume approaches zero. And since density is equal to mass/volume, the
numerator of the fraction remains the same, but the denominator approaches
zero. Thus the density approaches infinity.

6.3.2 Schwarzschild radius

Most people have heard the term "event horizon" of a black hole. The event
horizon is the point of no return, which is why Shep Doeleman of MIT calls it
"an exit door from the Universe." Nothing that approaches a black hole closer
than this distance can ever escape from the gravitational force of the black
hole. Even light cannot escape, which is why the object is called "black." In
this section, you will learn how to calculate the size of the event horizon or
"Schwarzschild radius" of a black hole.

   As you work through this section, keep in mind that the event horizon is a
mathematically defined distance from the center of the black hole. It describes
an imaginary spherical surface surrounding the central singularity, not a real
physical surface. The event horizon itself cannot be seen, and if you were
falling into a black hole, you may not know when you crossed the event
horizon. But once inside, you would have no hope of ever getting out.

   It's important to realize that when astronomers refer to the "size" of a
black hole, they are referring not to the zero volume that the mass physically

occupies, but rather to the size of the Schwarzschild radius (R_s). That radius is nonzero and can be determined by finding the distance at which the v_{esc} equals the speed of light. As you can see in the next section, this results in the following equation:

$$R_s = \frac{2Gm}{c^2}. \tag{6.6}$$

Although not a physical surface, the size of the event horizon is mathematically defined by the equation above, where R_s is shorthand for Schwarzschild radius, G is the universal gravitational constant, m is the mass comprising the singularity, and c is the constant speed of light.

The size of the event horizon depends on exactly one physical property of the black hole. Can you name it? If not, take a look at the parameters on the right side of Eq. 6.6. Notice that they're all constants, with one exception: the variable (m) that represents the mass of the black hole. Mass is the one and only physical property of a black hole that determines its event horizon. In this sense, black holes are actually simpler to analyze than other astronomical objects.

Example: The lowest-mass stellar black holes have masses of approximately 3 solar masses. How big are their event horizons?

You are given the mass ($m = 3\,\mathrm{M_\odot}$), and asked to calculate the Schwarzschild radius (R_s), which requires plugging values into Eq. 6.6:

$$R_s = \frac{2Gm}{c^2} = \frac{2G(3\,\mathrm{M_\odot})}{c^2} = \frac{2(6.67 \times 10^{-11}\,\mathrm{N m^2/kg^2})(3(1.99 \times 10^{30}\,\mathrm{kg}))}{(3 \times 10^8\,\mathrm{m/s})^2}$$

$$= 8.8 \times 10^3\,\mathrm{N} \cdot \frac{\frac{1}{\mathrm{kg}}}{\frac{1}{\mathrm{s^2}}}.$$

And, since the units of newtons (N) are equivalent to kg·m/s² as described in Section 1.1.6,

$$R_s = 8.8 \times 10^3 \frac{\mathrm{kg \cdot m}}{\mathrm{s^2}} \cdot \frac{\frac{1}{\mathrm{kg}}}{\frac{1}{\mathrm{s^2}}} = 8{,}800\,\mathrm{m} = 8.8\,\mathrm{km}.$$

Thus the event horizons of the smallest stellar black holes are just under 9 km from the singularity at the center. Compared to other astronomical objects, black-hole event horizons are quite small – about the size of a small city. Within this distance, nothing can escape from the black hole. But outside the event horizon, objects can orbit a black hole exactly as they would orbit any other object of the same mass. So, if the Sun were suddenly replaced with a

1-solar-mass black hole, the Earth and all the other planets would continue orbiting just as they currently do because they would all be safely outside the event horizon.

How much does the radius of the event horizon change as a black hole gains mass? To answer this question, take another look at Eq. 6.6. Notice that R_s is directly proportional to m, so if the black hole's mass increases by some factor, the radius of its event horizon grows by the same factor. This has the interesting implication that black holes can grow. Any matter or energy that passes through the event horizon contributes more mass to the black hole, which makes the event horizon larger. And if two equal-mass black holes merge, the result will be a black hole with twice the mass, and the event horizon of that black hole will be twice as large.

Example: Consider two black holes with different mass, $3\,M_\odot$ and $9\,M_\odot$. How do the radii of their event horizons compare?

Since R_s and m are directly proportional, comparison problems are particularly straightforward. If one black hole's mass is larger or smaller than another black hole's mass by some factor, its Schwarzschild radius will be larger or smaller by the same factor. Applying this reasoning to the $3\,M_\odot$ and $9\,M_\odot$ case, you know that the smaller black hole is one-third the mass of the larger one. Therefore, since their radii must share the same relationship, you can conclude that the smaller black hole's R_s must be one-third that of the larger one. Writing out all the math, starting with the desired ratio of radii, using Eq. 6.6, and plugging in the information given in the question, gives

$$\frac{R_{s,small}}{R_{s,large}} = \frac{2Gm_{small}/c^2}{2Gm_{large}/c^2} = \frac{2G(3\,M_\odot/c^2)}{2G(9\,M_\odot/c^2)} = \frac{3}{9} = \frac{1}{3},$$

or

$$R_{s,small} = \frac{1}{3}R_{s,large}.$$

This result agrees with the expected conclusion that the $3\,M_\odot$ black hole has one-third the Schwarzschild radius of the $9\,M_\odot$ black hole.

Although stellar black holes don't form in nature unless the remnant core has a mass of $3\,M_\odot$ or more, it's instructive to consider smaller masses in order to gain more insight into the mathematical relationship between R_s and m.

Example: How much would the Earth (6×10^{24} kg) have to be compressed for its radius to equal the Schwarzschild radius for that mass?

Plugging Earth's mass into Eq. 6.6 gives a Schwarzschild radius of

$$R_s = \frac{2Gm}{c^2} = \frac{2(6.67 \times 10^{-11} \text{N}\text{m}^2/\text{kg}^2)(3(6 \times 10^{24} \text{kg}))}{(3 \times 10^8 \text{m/s})^2}$$
$$= 8.9 \times 10^{-3} \text{ m},$$

which means that compressing the Earth to a tiny sphere of radius 8.9 mm would cause it to contract under its own gravity to form a black hole. But keep in mind that the Moon, being well outside the event horizon, would continue orbiting normally.

The following exercise will give you a sense of how far matter must be compressed to form a black hole, bearing in mind that black holes under 3 $M_\odot$ do not form in nature.

Exercise 6.7. Find the Schwarzschild radius for these objects:

(a) **A sphere with a mass of 1 kg.**
(b) **The Sun (try the ratio method).**
(c) **An object with mass of four million solar masses (you can learn more about this kind of black hole in the problems at the end of this chapter).**

6.3.3 Escape speed near black holes

You may be wondering how escape speed applies to black holes, since it is frequently claimed that "nothing can escape from a black hole, not even light." This statement is true if properly qualified: "nothing can escape from within the event horizon of a black hole." As mentioned in the previous section, there is nothing special about a black hole's gravitational influence; its gravitational force on any object and the v_{esc} required to escape from it are calculated just as they are for any other mass. That means you can use Eq. 2.1 to find the force of gravity between a black hole and another mass, and you can use Eq. 6.5 to find v_{esc} at a given distance from the center of a black hole. The fact that the mass of a black hole is compressed into a singularity has no bearing on these calculations, which apply both outside and inside the event horizon. So what is the difference between a black hole and any other object with the same mass? Only this: because a black hole has zero physical size, you can get arbitrarily close to all of its mass.

To understand why this is very different for other objects, consider how close you can get to all the mass of an object such as the Sun. Even if you were to fly your spaceship to within 1 meter of some point on the Sun's photosphere (effective "surface"), you'd still be over a million kilometers away from the material on the other side of the Sun, because the diameter of the Sun is about 1.4 million kilometers. But if you get within 1 meter of the singularity of a black hole, you're within 1 meter of the entire mass of that black hole. So it's not the total mass of a black hole that makes it dangerous, it's the concentration of that mass into a single point, since that allows you to get close to all of it simultaneously.

To escape from the gravitational pull of a black hole or any other object with mass m simply requires that you achieve a speed equal to or greater than v_{esc}, as defined in Section 6.2:

$$v_{esc} = \sqrt{\frac{2Gm}{R}}, \tag{6.5}$$

where R is the distance from the center of the mass from which you're trying to escape and m is the amount of that mass.

Now consider what would happen to v_{esc} as you get closer and closer to the central singularity of a black hole. To understand this, think about the limiting case of Eq. 6.5 as R approaches zero. In that case, v_{esc} approaches $\sqrt{\frac{2Gm}{0}}$, which means that the escape speed approaches infinity. And since nothing – not even light – can travel at infinite speed, not even light can escape.

So what is the significance of the event horizon? Escape speed increases steadily as you get closer and closer to a black hole, both inside and outside the event horizon. To find out what happens at the exact distance of the event horizon from the singularity, just use the expression for the Schwarzschild radius (R_s) from Eq. 6.6 as the distance (R) in Eq. 6.5:

$$v_{esc,\ event\ horizon} = \sqrt{\frac{2Gm}{R_s}} = \sqrt{\frac{2Gm}{\frac{2Gm}{c^2}}} = \sqrt{\frac{2Gm}{\frac{2Gm}{c^2}}} = \sqrt{c^2} = c.$$

This means that at the event horizon, v_{esc} is precisely the speed of light. In fact, this is how Schwarzschild radius is defined in the first place: the distance at which escape speed equals the speed of light. Outside the event horizon, R (distance) is larger so v_{esc} is smaller than c, and escape is possible. Inside the event horizon, v_{esc} is larger than c, which is the reason why nothing, not even light, can escape from this region.

Figure 6.3 shows how escape speed varies with distance from a black hole. In this graph, the vertical axis represents escape speed (v_{esc}), and the horizontal axis represents the distance (R) from the singularity. At very small distances (approaching the singularity, as $R \to 0$) escape speed approaches infinity ($v_{esc} \to \infty$), and at very large distances (as $R \to \infty$) escape speed drops off toward zero ($v_{esc} \to 0$). Thus the curve asymptotically approaches

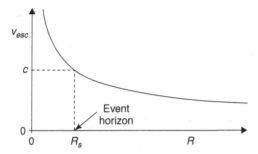

Figure 6.3 Decrease of escape speed with distance from singularity.

both axes but never reaches either one. At the Schwarzschild radius ($R = R_s$), which is a relatively small radius, the escape speed equals the speed of light ($v_{esc} = c$) by definition.

Exercise 6.8. Calculate the escape speed at the event horizon ($R = R_s$) for a 1 M$_\odot$ and a 10 M$_\odot$ black hole.

Exercise 6.9. How does v_{esc} at a distance of $4R_s$ compare to v_{esc} at R_s for each of the black holes of the previous exercise?

6.4 The expansion of the Universe

This section marks a transition from the smallest to the largest scales in astronomy – from individual black holes, which occupy zero volume, to cosmology, which is the study of all the matter and energy occupying the entire Universe.

One of the key discoveries in the history of cosmology is that all distant galaxies are moving away from our Milky Way galaxy, which leads to the conclusion that the Universe is expanding. Even more profound than the expansion itself is the realization that there is nothing special about our perspective on this expansion; any hypothetical observer, anywhere else in the Universe, would make the same measurement, which means that there is no unique center or edge to the Universe. Space is stretching everywhere, so all galaxies are moving away from all other galaxies, provided that they are not close enough together to be gravitationally bound to one another.

The Universe is expanding because empty space is stretching uniformly everywhere. This means that every bit of empty space stretches as much as every other bit. The amount of space between us and a distant galaxy, which is to say its distance from us, directly determines how fast we perceive it to be moving away from us, which is called its "recession speed." For example,

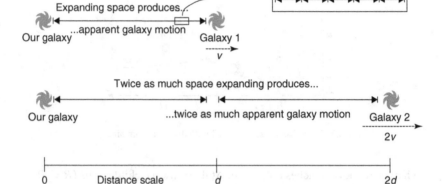

Figure 6.4 Two galaxies at different distances from our galaxy will have different velocities from our perspective.

if the distance from us to a certain galaxy is twice as far as the distance from us to another galaxy, there is twice as much expanding space between us and the more-distant galaxy, so it will appear to have twice the recession speed. Figure 6.4 illustrates this with a simple one-dimensional universe containing three galaxies.

If you're uncertain about why greater distance means greater velocity, it may help to put specific numbers into a thought experiment. Imagine that galaxy 1 is initially at distance d and galaxy 2 is initially at distance $2d$. Now imagine that over some interval of time t, the expansion of the Universe causes the space between the galaxies to double. That means that in that time, galaxy 1 will move from distance d to distance $2d$, so the change in galaxy 1's distance will be $2d - d = d$. But in that same amount of time, galaxy 2 will move from distance $2d$ to distance $4d$, so the change in galaxy 2's distance will be $4d - 2d = 2d$. Since speed equals distance divided by time, and since galaxy 2 is moving twice the distance in the same amount of time, the recession speed of galaxy 2 (speed $= 2d/t$) will be twice that of galaxy 1 (speed $= d/t$). This analysis works not only for the case of galaxy 2 twice as far away and space doubling – you could have chosen galaxies at any two distances and considered any expansion factor, and the velocity ratio will always be the same as the distance ratio.

At first blush, it appears that the relationship between distance and recession speed implies that the Milky Way galaxy is the center of the Universe. After all, if distance from *us* is the sole determining factor of recession speed, doesn't that mean that our position is somehow special?

You can see the fallacy of that argument by imaging yourself on galaxy 2 in Figure 6.4. The same analysis can be applied to this situation, and the results are symmetric: viewed from galaxy 2, the distance to galaxy 1 is d and the distance to our galaxy is $2d$, and as space expands, our galaxy is receding from galaxy 2 twice as fast as galaxy 1 is receding from galaxy 2. So an observer in galaxy 2 has equal right to conclude that distance from galaxy 2 is the sole determining factor of recession speed.

A similar analysis applies to a hypothetical observer in galaxy 1, who sees both our galaxy and galaxy 2 at distance d, albeit in opposite directions, and who measures the speed of recession of our galaxy and galaxy 2 to be equal. If our picture had included other galaxies to the left of our galaxy and to the right of galaxy 2, an observer on galaxy 1 would measure the recession speed of those galaxies to be greater than the recession speed of our galaxy and galaxy 2, because their distance from galaxy 1 would be greater.

In the real Universe, galaxies are not in a line, but are scattered throughout three-dimensional space. But the same simple proportionality holds true for galaxies receding from us in all directions.

The important conclusion is that the expansion of space means that no particular location is special – the same rule applies to every observer, and that rule is that more-distant galaxies are moving away more quickly. This is the essence of Hubble's Law, and you can see how to quantify this law in the next subsection.

Exercise 6.10. Consider another galaxy in Figure 6.4 that is initially a distance $d/2$ from our galaxy. How does its recession speed from us compare to that of galaxy 2 over the time interval t in which distances double? What about over a time interval in which distances triple?

6.4.1 The Hubble diagram and Hubble's Law

One of the most useful graphs in any discussion of the expansion of the Universe is a "Hubble diagram," sometimes called a "Hubble plot." Although the units may vary between texts, a Hubble diagram is typically a graph of the distance of many galaxies from us versus their recession speed. Figure 6.5 shows an example Hubble diagram with the distance and recession speed of 100 hypothetical galaxies, where each point represents one galaxy. This graph also includes a line of best fit through the points. Keep in mind that "diagram" in this context means a graph, not a picture, much like the H–R diagram described in Section 5.4.

There are several important features to notice on this sample Hubble diagram. First, all galaxies are moving away from us so all recession velocities

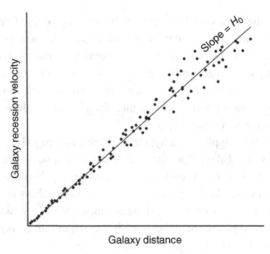

Figure 6.5 A standard Hubble diagram has speed of galaxies moving away from us on the y-axis, and distance away from us on the x-axis.

are positive, following the sign convention from Section 3.4 on Doppler shift in which positive velocity corresponds to motion away from the observer. The more striking feature, however, is the clear upward linear trend in the graph. This trend is a graphical representation of Hubble's Law as mentioned earlier in this section: more distant galaxies are receding from us more quickly.

You can understand the equation for the linear relationship between recession velocity and distance by recalling the equation of a straight line on a graph: $y = mx + b$, where b is the y-intercept (the point at which the line crosses the y-axis) and m is the slope. Since the Hubble diagram line goes through the origin, the y-intercept is zero ($b = 0$) and the equation simplifies to $y = mx$. In the Hubble diagram the y-values are velocity v, the x-values are distance d, and the slope is called the Hubble constant, or H_0, pronounced "H-naught":

$$v = H_0 d. \tag{6.7}$$

This equation is the classic mathematical statement of Hubble's Law. It says that a galaxy's recession velocity and distance from us are directly proportional, with the constant of proportionality of H_0. This is consistent with the discussion of Figure 6.4, in which galaxy 2 was twice as distant as galaxy 1 and therefore showed a recession velocity that was twice as large.

You may be wondering why the points are scattered around the line, given that the equation for Hubble's Law is a perfect linear relationship with no scatter. One cause of the scatter is measurement uncertainty – there are uncertainties in every measurement made in science. These uncertainties are often shown on graphs as "error bars," but that doesn't mean that an error was made. Error bars represent an acknowledgment of the finite precision of the measurement, akin to specifying how many significant figures can be trusted in your answer. Careful analysis of measurement uncertainty is one of the hallmarks of good science.

Another cause of scatter in the points is galaxy motion due to the gravitational forces that galaxies exert upon one another. These forces can produce "local" galaxy motions that affect the recession speed we measure, which means that not all of the points on the graph will fall on the same line even if the measurements have very high precision.

Exercise 6.11. Sketch a Hubble diagram for a uniformly *contracting* universe in which all galaxies are moving toward us with a speed proportional to their distance.

6.4.2 The numerical value of H_0

How can you determine the value of the Hubble constant H_0? One approach is to obtain it directly from the Hubble graph by calculating the slope of the line. To do this, you need a Hubble diagram with numerical values on the axes, as shown in Figure 6.6. Remember that the slope of a line is a measure of how steep it is: positive slopes go upward from left to right, and the larger the slope,

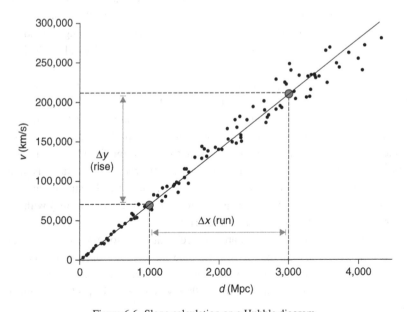

Figure 6.6 Slope calculation on a Hubble diagram.

the more steeply the line goes up. Mathematically, slope (m) is defined as the "rise" (the change in y-values, called Δy, over some interval) divided by the "run" (the change in x-values, called Δx, over the same interval):

$$m = \frac{\Delta y}{\Delta x}. \tag{6.8}$$

For the rise and run depicted in Figure 6.6, the slope calculation is:

$$m = \frac{\Delta y}{\Delta x} = \frac{210{,}000 \text{ km/s} - 70{,}000 \text{ km/s}}{3{,}000 \text{ Mpc} - 1{,}000 \text{ Mpc}} = \frac{140{,}000 \text{ km/s}}{2{,}000 \text{ Mpc}} = \frac{140 \text{ km/s}}{2 \text{ Mpc}},$$

$$H_0 = 70 \text{ (km/s)/Mpc}.$$

Here are some tips for calculating slope. First, the order of the subtraction does not matter; reversing the order in the example above gives

$$m = \frac{70{,}000 \text{ km/s} - 210{,}000 \text{ km/s}}{1{,}000 \text{ Mpc} - 3{,}000 \text{ Mpc}} = \frac{-140{,}000 \text{ km/s}}{-2{,}000 \text{ Mpc}} = 70 \text{ (km/s)/Mpc}.$$

The negative signs cancel, so the answer is the same.

Second, be sure to keep the units with the numbers. The constant H_0 has dimensions that are physically meaningful (speed per distance), and the units must be retained in order to do calculations with this constant later.

Third, it doesn't matter which two points you choose, as long as they are on the line. Since the line is straight, the slope is constant everywhere on it. The points don't have to be galaxies from the graph at all; they can be two arbitrary points on the line. In fact you are better off *not* choosing individual galaxy points from the graph, since most of them don't actually fall on the line itself but are scattered around it.

Fourth, if you are not provided with a line of best fit, you'll have to create one for yourself. It should be a perfectly straight line going roughly through the middle of the clump of points and the origin. Excepting for distant outliers, roughly half the points should fall above your line, and half below. If you're sketching it on paper, you are probably safe to eyeball the fit unless instructed otherwise, but be sure to use a straightedge. If you're making a graph using a computer program, you can use a line-fitting function; just be sure to specify the fitting function as a first-order polynomial (that is, a straight line) with no offset to ensure that it goes through the origin.

Fifth, try to choose points that are relatively far apart to minimize the effect of uncertainties. For example, if you can read the y-values only to the nearest 10,000 km/s, this uncertainty has a greater effect on the value of the slope when divided by a small number for Δx than when divided by a large number for Δx.

Finally, since you are free to choose your own points on the line, it is convenient to choose points at nice round numbers (such as the values of 1,000 Mpc and 3,000 Mpc in the example) in order to make your subtraction simpler. In fact if your line goes through the origin, which it always should in a Hubble diagram, then $(0, 0)$ is a particularly convenient choice for one of your points. Using the origin for the lower left point in the slope calculation above (that is, using $x = 0$, $y = 0$ instead of $x = 1,000$, $y = 70,000$) would have simplified the calculation to

$$\frac{\Delta y}{\Delta x} = \frac{210,000 \text{ km/s} - 0 \text{ km/s}}{3,000 \text{ Mpc} - 0 \text{ Mpc}} = \frac{210,000 \text{ km/s}}{3,000 \text{ Mpc}} = 70 \text{ (km/s)/Mpc}.$$

This agrees with the value for H_0 as found previously, because you have calculated the slope of the same line using different points.

Calculations of the value of H_0 from many different astronomical observations in recent decades have given results that range from the high 60s to the mid 70s, in units of (km/s)/Mpc. Some books and articles give the value of H_0 in (km/s)/Mly (km/s per million light years), in which case the value falls in the range of 21–23 (km/s)/Mly. The Universe has only one true Hubble constant at any time, but the inherent uncertainty of our measurements results in a range of values.

As mentioned in the previous section, the variations in measured values for H_0 can also be caused by variations in the recession speed of individual galaxies due to gravitational interactions with other galaxies. Corrections have been made for known gravitational effects, and hundreds of measurements have been made to determine the value of the Hubble constant. In recent decades, those measurements have "converged" – that is, they have clustered together within a small range. That range is about 67–72 (km/s)/Mpc, and the true value of H_0 is very likely to be somewhere near the middle of that range. If you're working a Hubble problem and you're not given a value of H_0 to use in your calculations, you should be safe using a value in that range. Some of the calculations in this section use a rounded value of H_0 of 70 (km/s)/Mpc for simplicity, but keep in mind that other texts or your professor might specify a slightly different value.

Exercise 6.12. In Figure 6.6, find the value of H_0, the slope of the line, using two points on the line at $x = 500$ and $x = 4,000$ Mpc.

Exercise 6.13. Repeat the calculation for the slope using the origin $(0, 0)$ instead of $x = 500$ Mpc for the lower point.

Exercise 6.14. Reverse the order of subtraction and repeat the previous calculation for the slope.

6.4.3 Calculations with Hubble's Law

Armed with a value for the Hubble constant, you can use Hubble's Law (Eq. 6.7) to calculate the recession speed of a distant galaxy if you're given its distance, or you can calculate its distance of you're given its recession speed. In actual practice, astronomers measure the recession speed of distant galaxies using the Doppler effect, which makes Hubble's Law a very powerful distance-measuring tool.

You should be aware that Hubble's Law is not useful for nearby galaxies, within about 3 Mpc of us, because galaxies within our "local group" are gravitationally bound to our own Milky Way galaxy and the Andromeda Galaxy. These strong gravitational links overcome the expansion of the Universe, so nearby galaxies do not move away from us at the expansion rate of the Universe, even though the space between us is expanding.

When you're doing calculations using Hubble's Law, you should be sure to retain the units of H_0. If the units of the distance or recession speed you are given are not consistent with the corresponding units in H_0, you must perform a unit conversion, as in the following example.

Example: If a galaxy is two billion parsecs away, what is its recession speed due to the expansion of the Universe?

According to Hubble's Law (Eq. 6.7), a galaxy's recession speed is equal to its distance from us times the Hubble constant. Since you were not told what specific value to use for H_0, a safe bet is to use the generally accepted round value of 70 (km/s)/Mpc. Use 2,000 Mpc for d because 2,000 million is the same as two billion. Plugging in H_0 and d gives you

$$v = H_0 d = (70 \text{ (km/s)/Mpc})(2,000 \text{ Mpc}) = 140,000 \text{ km/s}.$$

Because this galaxy is very far away (two billion parsecs is almost seven billion light years), it is moving very fast – almost half the speed of light. This was a straightforward "plug-and-chug" problem using the equation for Hubble's Law, since you were given the distance in units that were easily converted to megaparsecs. Notice that explicitly carrying the units through with the numbers makes it clear that the answer comes out in units of kilometers per second.

You could also have solved this problem by looking at a Hubble diagram whose slope is equal to the value of H_0, such as Figure 6.6. By finding the value $x = 2,000$ Mpc on the x-axis, moving straight up to the best-fit line, and then moving left to the y-axis, you can read the corresponding y-value of 140,000 km/s.

Example: If a Doppler-shift measurement shows that a certain galaxy is receding from us at a speed of 10,000 km/s, how far away is that galaxy?

This time you are asked to calculate the distance, so you must first rearrange Eq. 6.7 to solve for distance:

$$d = \frac{v}{H_0}. \tag{6.9}$$

Then plugging in the values for v (given in the problem statement) and H_0 (assuming the standard value) gives

$$d = \frac{10,000 \text{ km/s}}{70 \frac{\text{km/s}}{\text{Mpc}}} = \frac{1,000}{7} \text{ Mpc} \approx 143 \text{ Mpc},$$

or 143×10^6 pc. If you wanted to express this answer in light years instead of parsecs, you could do a unit conversion:

$$143 \times 10^6 \text{ pc} \left(\frac{3.26 \text{ ly}}{1 \text{ pc}} \right) = 466 \times 10^6 \text{ ly} = 466 \text{ Mly},$$

or almost half a billion light years.

You can also solve this problem by reading the x-value from the graph in Figure 6.6, as described in the previous example. Or, if you already have a pair of numbers that you know are related by Hubble's Law such as $v = 140,000$ km/s and $d = 2,000$ Mpc from the previous example, you can use the ratio method to do this problem. Writing out the Hubble's Law equation twice and dividing them, the constant will cancel:

$$\frac{v_2 = H_0 d_2}{v_1 = H_0 d_1} \quad \Rightarrow \quad \frac{v_2}{v_1} = \frac{H_0 d_2}{H_0 d_1} \quad \Rightarrow \quad \frac{v_2}{v_1} = \frac{d_2}{d_1}.$$

Now solving for the desired quantity – the distance of galaxy 2 – and then plugging in the values of v and d from the previous example for galaxy 1, and the velocity given in this problem for galaxy 2, you get

$$d_2 = d_1 \left(\frac{v_2}{v_1} \right) = 2,000 \text{ Mpc} \left(\frac{10,000 \text{ km/s}}{140,000 \text{ km/s}} \right) = \frac{2,000}{14} \text{ Mpc} \approx 143 \text{ Mpc},$$

which is in agreement with the previous result. The ratio method took about as many mathematical steps as the absolute method to solve this problem, but it did not require knowing the value of H_0.

Exercise 6.15. If a galaxy is receding from us with a speed of 15,000 km/s, how far away is that galaxy?

6.4.4 The age of the Universe

Since all distant galaxies are moving apart due to the expansion of the Universe, it's logical to conclude that in the past those galaxies were all closer together. Earlier in the past, they were closer still. Extrapolating this expansion backward, at some time in the past all galaxies (or the material of which they are now composed) were in the same place. That is, the space between all matter in the Universe was zero, and the Universe occupied zero volume. That instant when the Universe began expanding is called the Big Bang. How long ago did that happen?

Hubble's Law allows you to calculate, or at least estimate, the age of the Universe – that is, the amount of time that has passed since the Universe began expanding. If you know how fast a galaxy is moving away from you, and you also know how far it has travelled since you were together, you can determine how much time has passed since that galaxy was at your location. To do this, use Eq. 1.11:

$$\text{time} = \frac{\text{distance}}{\text{speed}}. \tag{1.11}$$

Applying this relationship to distance and recession speed of galaxies due to the expansion of the Universe, you can use v from Hubble's Law ($v = H_0 d$) for speed in Eq. 1.11 to give

$$\text{time} = \frac{\text{distance}}{\text{speed}} = \frac{\text{distance}}{H_0 \times \text{distance}} = \frac{1}{H_0}.$$

The result is an estimate of the age of the Universe assuming that the rate of expansion has been constant. This age is often called T_0 or the "Hubble time," and is related to the Hubble constant by

$$T_0 = \frac{1}{H_0}. \tag{6.10}$$

If you are careful about keeping track of units, plugging in a value for H_0 gives you a numerical value of the age of the Universe. Notice that this age does not depend on v or d for any individual galaxy, because the age of the Universe is not specific to any one galaxy.

Example: What is the age of the Universe if $H_0 = 70$ (km/s)/Mpc?

Plugging in 70 (km/s)/Mpc for the Hubble constant in Eq. 6.10 gives

$$T_0 = \frac{1}{70 \, \frac{\text{km/s}}{\text{Mpc}}} = \frac{1 \, \text{Mpc}}{70 \, \text{km/s}} = \frac{10^6 \, \text{pc}}{70 \, \text{km/s}}.$$

In order to cancel the distance units in the numerator (pc) with the distance units in the denominator (km), you must do a unit conversion between parsecs and kilometers. If you look up the equivalence between these units (1 pc ↔ 3.09×10^{13} km), you can do this with one conversion factor:

$$T_0 = \frac{10^6 \text{ pc}}{70 \text{ km/s}} \left(\frac{3.09 \times 10^{13} \text{ km}}{1 \text{ pc}} \right) = \frac{3.09 \times 10^{19}}{70 \frac{1}{\text{s}}} \approx 4.41 \times 10^{17} \text{ s}.$$

Converting this result to years may give you a better sense of how long this is. The conversion from seconds to years can be tacked on to the conversion between parsecs and kilometers above, or it can be performed separately:

$$T_0 = 4.41 \times 10^{17} \text{ s} \left(\frac{1 \text{ yr}}{3.16 \times 10^7 \text{ s}} \right) \approx 14 \times 10^9 \text{ yr},$$

or about 14 Gyr. This is close to the widely accepted value of 13.7 billion years for the age of the Universe, so T_0 gives a reasonable estimate.

Exercise 6.16. Calculate the Hubble time (T_0) for constant expansion if the value of H_0 were 69 (km/s)/Mpc. What if H_0 were instead 75 (km/s)/Mpc?

This approach to determining the age of the Universe is equivalent to spotting a person at a distance of 8 meters moving directly away from you at a speed of 1 meter per second and inferring how long since they were at your location. Based on your observations, you might conclude that the person has been moving away from you for a period of 8 seconds, since distance equals speed times time, and 8 meters = 1 meter per second × 8 seconds. But that conclusion contains an important assumption that the person's speed has been the same over those 8 seconds.

The assumption of constant speed is built right into the equation distance = speed × time, which is only a portion of the full equation relating distance (d) to time (t). You might have seen that full equation if you've ever taken an introductory physics class. It is

$$d = \frac{1}{2}at^2 + v_0 t, \tag{6.11}$$

in which a represents acceleration and v_0 is the initial speed of the object. So every time you set distance equal to speed × time, remember that you're ignoring the first term of Eq. 6.11, which means that you're assuming zero acceleration, so the speed of the object is not changing.

Thus the technique of finding the age of the Universe (T_0) by taking the reciprocal of the Hubble constant ($1/H_0$) is based on the assumption that the rate of expansion of the Universe has not changed since the Big Bang. But ever since Georges Lemaître and Edwin Hubble discovered the expansion of the Universe in the late 1920s, astronomers have suspected that the expansion rate is not constant. For several decades, many believed that the acceleration should be negative and the rate of expansion slowing due to the pull of gravity of all the galaxies upon one another. But in the late 1990s, astronomers made an astonishing discovery: the rate of the expansion of the Universe is not slowing down, it's speeding up. The source of this positive acceleration is not fully understood (it's called "dark energy" for lack of a better term), but Doppler-shift data from distant supernovae indicate that the rate of expansion was slower a few billion years ago than it is today.

To understand the effect of the changing expansion rate on the age and fate of the Universe, you should first understand a type of graph that's somewhat different from the Hubble graph. This type of graph is called a "position-vs.-time" graph, and it's the subject of the next section.

6.4.5 Position-vs.-time graphs

One easy way to understand position-vs.-time graphs is to imagine a graph showing the position of a friend who's walking away from you or toward you over time. Such a graph is shown in Figure 6.7, with time increasing to the right on the horizontal axis and your friend's position increasing upward on the vertical axis. In this type of graph, you're at position zero, so your friend's position value is her distance from you.

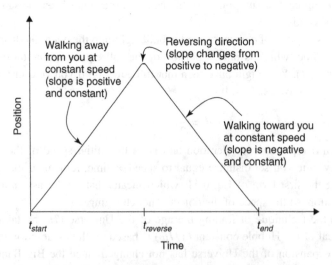

Figure 6.7 Position-vs.-time graph for friend walking away and back.

In the scenario shown in this figure, your friend starts out at your position (zero height along the vertical axis) at time $t_{start} = 0$ and initially walks away from you at constant speed. Since her speed is constant, her position increases linearly (that is, along a straight line on the graph) over time. That's because for constant speed, distance equals speed × time, and with distance (d) on the y-axis and time (t) on the x-axis, $d = vt$ is the equation of a straight line. As you may recall from Section 6.4.1, the equation for a straight line is $y = mx + b$, where m is the slope of the line and b is the y-intercept. The y-intercept is zero in this case since your friend started out at your position at time $t = 0$, so the equation of this line can be written as $y = mx$. Comparing $d = vt$ to $y = mx$, you can see another important relationship: your friend's speed (v) is equal to the slope of the line (m) on her position-vs.-time graph. That should make sense to you, since the slope of a line is defined as the rise (Δy) over the run (Δx), and in this case the rise is the change in position (distance) while the run is the time it takes for that position change. Hence on a position-vs.-time graph

$$\text{slope} = \frac{\Delta y}{\Delta x} = \frac{\text{distance}}{\text{time}} = \text{speed}, \tag{6.12}$$

where positive slope means increasing distance.

In the scenario shown in Figure 6.7, after walking away for some time, your friend turns around and walks back toward your location at the same constant speed with which she walked away. This part of her journey is shown by the portion of the graph in which her position value gets smaller over time, since smaller position means less distance between you and your friend. For the return trip, the slope of your friend's line is again constant, since she's walking at a constant speed, but now the slope of her line will be negative, since the change in her position (Δy) is negative when she's moving toward you. And since she walks back at the same speed and for the same amount of time as she walked away, your friend ends up back at your position at the end of her trip.

There's one portion of the graph in Figure 6.7 in which the slope of the line is not constant. That's the portion near the time ($t_{reverse}$) at which she changes her direction from moving away to moving toward you. As she approaches the turn-around point, she slows down, making the slope of her line less positive. At the instant she stops moving away, her slope momentarily equals zero, and as she begins moving toward you, her slope becomes negative – slightly at first, and then reaching a constant value as she gets up to her walking speed.

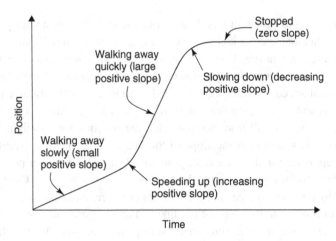

Figure 6.8 Position-vs.-time graph for varying speeds.

Example: Sketch a graph summarizing the position-vs.-time motion if your friend begins walking away from your position at a slow speed but then speeds up and walks away more quickly; after moving away at this higher speed for a short time, she slows down, comes to a halt, and remains stationary at her position for the remainder of the graph.

This example shows the effect of acceleration on position-vs.-time graphs. The details of this trip are annotated on the graph in Figure 6.8, and you should make sure you understand what's happening in each portion of the graph. But you should also take a step back and look at the big picture. In the big-picture view, straight lines mean constant speed, with a positive slope indicating away, a negative slope toward, and a zero (flat) slope indicating no motion. Curved lines have a changing slope and thus show a changing speed, which indicates acceleration: speeding up if the slope is becoming steeper (more positive *or* more negative) or slowing down if the slope is becoming flatter (less positive or less negative).

The use of position-vs.-time graphs in cosmology is discussed in the next section, but before moving on, you should work through the following exercise to verify your understanding of this important type of graph.

Exercise 6.17. Describe the motion of an object for which the position-vs.-time graph looks like the graph in Figure 6.9.

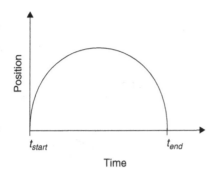

Figure 6.9 Position-vs.-time graph for Exercise 6.17.

6.5 The history and fate of the Universe

Some of the most fundamental questions in cosmology relate to the evolution of the Universe – how it began, how it became the Universe we see today, and how it will change in the future. In discussing these questions, many astronomy texts and articles show a version of the cosmological position-vs.-time graph shown in Figure 6.10. In this graph, four possible scenarios for the past and future expansion behavior of the Universe are presented. Observations of the expansion rate show that it slowed for a time in the past, and then began accelerating, so it pays to understand how to recognize both behaviors on the graph. This graph is useful because it conveys a great deal of information, and this section will be devoted to teaching you how to glean that information. By the end of this section, you should understand why the axes are labeled as they are, how the graph implies a Big Bang, how to read the age of the Universe from the graph, and what ultimate fate of the Universe is implied by each curve. Although this section does not contain many mathematical calculations, it emphasizes the widely applicable quantitative reasoning skill of reading and interpreting graphs.

6.5.1 Cosmological position-vs.-time graphs

If you compare the cosmological position-vs.-time graphs of this section to the generic position-vs.-time graphs of Section 6.4.5, you'll see two important differences. The first is that the vertical axis is placed near the middle of the graph rather than at the left edge. That's because cosmological position-vs.-time graphs often consider the present day to be "time zero" and place the vertical axis at that time, with the past to the left and the future to the right.

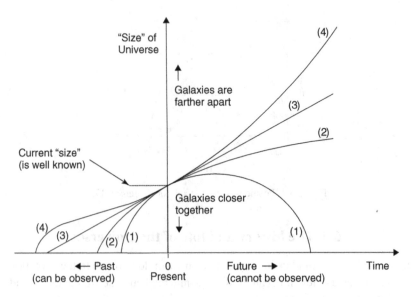

Figure 6.10 Axis labels on the graph of Universe expansion.

That way, the entire history and future of the Universe can be shown on the same graph. Figure 6.10 shows an annotated view of a cosmological position-vs.-time graph.

The other difference in cosmological position-vs.-time graphs is that the vertical axis is labeled "size" rather than position. The reason size is in quotes is that the size of the whole Universe is not known, and in fact may be unknowable. Observations can only probe the "observable Universe," which is the portion of space from which light has had time to reach the Earth. The entire Universe may be much larger, or even infinite in extent.

As a proxy for size on cosmological position-vs.-time graphs, the vertical axis actually represents the average spacing between galaxies, which is a measurable quantity. The average spacing between galaxies increases in the upward direction on these graphs, which means that galaxies are farther apart for points higher on the graph and closer together for points lower on the graph. Some astronomy texts use size as a label for the vertical axis because it is intuitive to picture an entire object growing or shrinking, and this graph represents the evolution of the whole Universe. Others may use the more-correct "galaxy separation" or similar label, based on actual measurements of average galaxy spacing.

Notice also that the curves for the four scenarios of the evolution of the Universe shown in Figure 6.10 converge to a point on the vertical axis at a

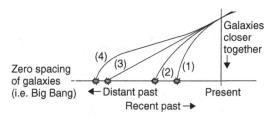

Figure 6.11 The instant of the Big Bang is the point at which each curve intersects the time axis.

time corresponding to the present day. That's because the present-day spacing of galaxies is known from observations, so any plausible scenario for the Universe's expansion must pass through this point.

So what kind of information can you find on graphs like this? Consider the points at which each of the four curves intersects the horizontal axis, as shown in Figure 6.11. Since the horizontal axis lies at the bottom of the vertical (size) axis, the space between galaxies is zero for all points on the horizontal axis. For these points, the Universe occupies zero volume and therefore has infinite density. So these points represent the instant of the Big Bang – the beginning of the expansion of the Universe.

Since each of the four curves begins with a Universe of zero size, all four scenarios start with a Big Bang. This is represented in Figure 6.11 by an explosion symbol at the appropriate time, but don't make the mistake of thinking of the Big Bang as an explosion in which matter and energy spread out into a pre-existing Universe, like a grenade exploding in an empty room. It's the Universe itself that's expanding, and there's no empty space into which it's expanding. It's difficult to picture, but there's nothing – not even a vacuum – outside the expanding Universe.

Exercise 6.18. **Specify which of the four curves in Figure 6.10 represents a Universe that will re-collapse to zero size in the future, ending with a reverse Big Bang (this is sometimes called the "Big Crunch" or "gnaB giB," which is Big Bang backwards).**

6.5.2 Determining the age of the Universe

In everyday language, the age of an object is defined as the time that has elapsed from its birth to the present time, and this same definition can be applied to the Universe. The previous section showed you how to identify the beginning of the expansion of the Universe in each of the four scenarios: find

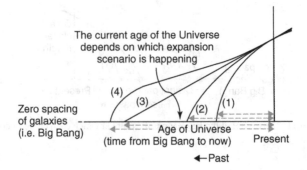

Figure 6.12 Age of the Universe for four expansion scenarios.

the time at which the curve intersects the time axis. So to determine the current age of the Universe, just identify the interval of time between that point and the present day which corresponds to the location of the vertical axis. That duration for each scenario is indicated by the dashed arrows in Figure 6.12.

Example: Which scenario implies the youngest age of the Universe?

Since the length of each dashed arrow in Figure 6.12 represents the current age of the Universe, you can answer this question simply by determining which scenario has the shortest arrow. This is scenario (1), in which the Big Bang occurred in the most recent past (closest to now).

Exercise 6.19. Rank the four scenarios shown in Figure 6.12 by the age of the Universe, from youngest to oldest.

If you put numerical labels on the axes of the graph, how would these four different ages of the Universe compare to T_0, the age calculated in Section 6.4.4? Remember that the earlier calculation was based on the assumption of constant expansion rate. This is consistent with the simplest of the four curves – scenario (3), represented by a straight line. But the other three scenarios each indicate an expansion rate of the Universe that has not been constant, and the implications of those past rate variations are discussed in the next section.

6.5.3 Changing past expansion rate

Just as in the generic position-vs.-time graphs discussed in Section 6.4.5, the slopes of the curves in the cosmological graphs in Figures 6.10–6.12 represent speed – in this case, the speed is the rate of expansion of the Universe. This rate can be positive (expansion), negative (contraction), or zero (constant size), and the steeper the slope (either positive or negative), the faster the expansion or contraction.

Example: Explain what the slopes of curves (1) and (3) imply about the past physical behavior of the Universe in those scenarios.

Figure 6.13 highlights the slopes for these two scenarios, allowing detailed analysis of their expansion behavior with time.

You already know the implication of the constant slope of scenario (3): the straight line represents a constant rate of expansion over all time. This means that in scenario (3) the expansion rate measured at the present time is the same as the expansion rate at all past times since the Big Bang, and at all future times.

Scenario (1) shows a changing slope, which indicates a changing expansion rate. After the Big Bang at the left extreme of curve (1), the initial slope was steeply upward – much steeper than the slope in curve (3). This means the Universe was expanding very quickly at first. But, immediately and smoothly, the slope began to grow shallower. This means that the expansion rate gradually slowed after the Big Bang and is continuing to slow. Note that the slope has been positive since the Big Bang and remains positive today, so the Universe has always been expanding, but it has been gradually slowing.

Exercise 6.20. Explain what the slopes of curves (2) and (4) imply about the physical behavior of the Universe in those scenarios.

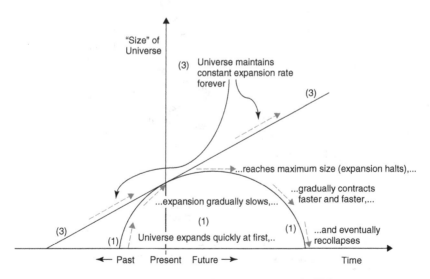

Figure 6.13 Relation of slope to expansion rate of the Universe.

6.5.4 Changing future expansion rate

Just as the shape of the curves to the left of the "present" time tell you how the expansion rate of the Universe behaved in the past, the curves to the right of present tell you what to expect in the future.

For example, look at the right portion of the curve for scenario (1) in Figure 6.13. Since the slope remains positive for some time after the present day, the Universe will continue to expand for that time. But eventually the slope becomes zero, which means that the expansion will grind to a halt, and the Universe will have reached a maximum size. The slope then turns negative, meaning that the Universe will begin contracting – slowly at first, then faster since the slope becomes more negative. Eventually, the distance between all the galaxies will be zero, and the Universe will return to its original size – occupying zero volume and infinitely dense.

The other scenarios predict very different ends for the Universe. In the other three scenarios, the slope of the curves never becomes negative, so the Universe does not shrink. Instead, the expansion continues forever and the Universe grows ever larger and less dense. Figure 6.14 highlights the end behavior of all four scenarios.

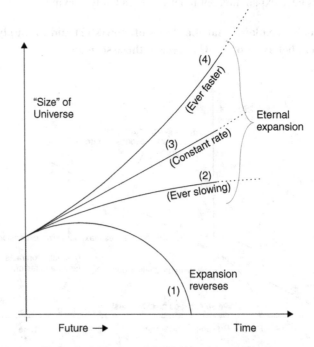

Figure 6.14 The fate of the Universe in each scenario.

Example: Under which scenario(s) will the Universe reach or approach a constant size?

As described in the preceding analysis, scenario (1) reaches a maximum size midway through its evolution when the slope of its curve becomes zero. However, this maximum size is momentary, and the Universe subsequently collapses. In scenario (2), the expansion of the Universe is perpetually slowing, but never quite stops. This behavior is an asymptote – the curve approaches zero slope and the Universe approaches a maximum size as time approaches infinity. This is analogous to the behavior of a projectile traveling at precisely the escape speed from another object: the Universe has just barely enough speed to escape from itself.

So only scenario (2) reaches a constant size. Notice that in the other two cases, (3) and (4), the Universe approaches an infinite size as time approaches infinity because it never slows down. In those scenarios, for any arbitrarily large size you choose, the Universe will eventually surpass it.

Exercise 6.21. Under which scenario(s) will the Universe reach or approach an infinite rate of expansion?

Exercise 6.22. Observational evidence shows that the expansion rate has accelerated in the past and is doing so now. If this acceleration continues in the future, to which scenario does that best correspond?

6.6 Chapter problems

6.1 Calculate the densities of the following objects: a white dwarf (same mass as Sun; same radius as Earth), a neutron star (three times Sun's mass, 1/1,000th of Earth's radius), and a black hole (same mass as the neutron star, zero radius).

6.2 Find the surface escape speeds of the objects in the previous problem.

6.3 When the Sun becomes a red giant, its radius will expand to approximately 1 AU, and its mass will remain approximately the same. By what factors will its density and escape speed change?

6.4 A spherical asteroid has a density of 2 g/cm^3 and a mass of 3×10^{19} kg. What is its radius?

6.5 If the escape speed from the surface of a certain planet is 5 km/s and the planet's density is 4,500 kg/m^3, what is the planet's radius?

6.6 How does the escape speed from the top of Mount Everest compare to the escape speed from the bottom of the Grand Canyon?

6.7 The highest-mass black holes known in nature are "supermassive" black holes found at the centers of galaxies. These black holes have mass of millions or even billions of solar masses. How big is the event horizon of a 1 billion solar-mass black hole in astronomical units?

6.8 Defining the "average density" of a black hole as its mass divided by the spherical volume within the Schwarzschild radius, find the average density of the supermassive black hole of the previous problem.

6.9 Instead of using the existing line of best fit drawn in Figure 6.6, imagine drawing your own line that goes through the origin and the galaxy point farthest *below* the existing line.

   (a) Do you expect this line to have a slope *larger, smaller*, or *the same* as the slope of the existing line? Explain your reasoning.

   (b) Using the origin (0, 0) and the x- and y-values of the "low" galaxy point you just selected, calculate the slope of your hypothetical line to verify or refute your prediction in the previous question.

6.10 Two galaxies have distances from Earth of 123 Mpc and 456 Mpc, respectively. Imagine you don't know the value of the Hubble constant (which is not 70 (km/s)/Mpc for this problem).

   (a) The closer galaxy has a recession speed of 9,594 km/s. Use the ratio method to calculate the recession speed of the other galaxy.

   (b) Calculate the value of the Hubble constant for this scenario.

   (c) If these were real galaxies in our Universe, explain why this value of H_0 would or would not surprise you.

6.11 A certain galaxy cluster (a large group of galaxies) has one trillion solar masses of material. Another galaxy cluster is 100 million light years away, receding due to the expansion of the Universe. How does the recession speed from the cluster due to the expansion compare to the escape speed that would be required for the second cluster to escape from the first one?

6.12 Compared to the Hubble time (T_0) estimate for the Universe's age, how would the actual age be different if the expansion had always been speeding up in the past? How would the actual age be different if the expansion had instead always been slowing in the past?

Further reading

Bennett, J., Donahue, M., and Schneider, N., *The Cosmic Perspective*, Addison-Wesley 2009.

Carroll, B. and Ostlie, D., *Astrophysics*, Addison-Wesley 2006.

Chaisson, E. and McMillan, S., *Astronomy Today*, Addison-Wesley 2010.

Dickinson, T., *The Backyard Astronomer's Guide*, Firefly 2008.

Freedman, R. and Kaufmann, W., *Universe*, W.H. Freeman 2007.

Hoskin, M., *The Cambridge Concise History of Astronomy*, Cambridge University Press 1999.

Mitton, J., *The Cambridge Illustrated Dictionary of Astronomy*, Cambridge University Press 2008.

Pasachoff, J. and Filippenko, A., *The Cosmos*, Brooks Cole 2006.

Schneider, S. and Arny, T., *Pathways to Astronomy*, McGraw-Hill 2011.

Seeds, M. and Backman, D., *Foundations of Astronomy*, Brooks Cole 2012.

Index

Printed in the United States
by Baker & Taylor Publisher Services